Landscape Approaches in Mammalian Ecology and Conservation

Landscape Approaches in Mammalian Ecology and Conservation

William Z. Lidicker Jr., editor

 University of Minnesota Press
Minneapolis
London

Published by the University of Minnesota Press
111 Third Avenue South, Suite 290
Minneapolis, MN 55401-2520

Printed in the United States of America on acid-free paper

Library of Congress Cataloging-in-Publication Data

Landscape approaches in mammalian ecology and conservation / William Z.
    Lidicker, Jr., editor.
        p.   cm.
    Based on papers from the Sixth International Theriological Congress, held
in Sydney, Australia, July 1993.
    Includes bibliographical references and index.
    ISBN 0-8166-2587-5 (hc)
    1. Mammals—Ecology—Congresses.   2. Landscape ecology—Congresses.
I. Lidicker, William Zander, 1932–   . II. International Theriological Congress
(6th : 1993 : Sydney, Australia)
QL.739.8.L36   1995
599.05—dc20   95-38448

The University of Minnesota is an equal-opportunity educator and employer.

Contents

Preface   vii

Part I.   Context: History, Theory, and Conservation

1. The Landscape Concept: Something Old, Something New
William Z. Lidicker Jr.   3

2. Development and Application of Landscape Approaches
in Mammalian Ecology
Lennart Hansson   20

Part II.   Field Approaches: Evidence and Contributions to Theory

3. Extinction and Survival of Rainforest Mammals in a Fragmented
Tropical Landscape
William F. Laurance   46

4. Movement in Spatially Divided Populations: Responses to Landscape
Structure
Gray Merriam   64

5. Movements of Small Mammals in a Landscape: Patch Restriction or
Nomadism?
Michał Kozakiewicz and Jakub Szacki   78

6. *Martes* Populations as Indicators of Habitat Spatial Patterns: The Need
for a Multiscale Approach
John A. Bissonette and Sim Broekhuizen   95

7. The Influence of Habitat Heterogeneity on Predator-Prey Dynamics
Tarja Oksanen and Michael Schneider   122

Part III.   Model Systems: An Experimental Protocol

8. Reflections on the Use of Experimental Landscapes in Mammalian
Ecology
Gary W. Barrett, John D. Peles, and Steven J. Harper   157

9. Population Dynamics of Small Mammals in Fragmented and Continuous
Old-Field Habitat
*James E. Diffendorfer, Norman A. Slade, Michael S. Gaines, and
Robert D. Holt*   175

Epilogue   200

Contributors   203

Author Index   205

Subject Index   211

Preface

Let the young in spirit look to the future and think landscape.
R. T. T. Forman and M. Godron, *Landscape Ecology*

Landscapes, the traditional grist for poets and artists, are increasingly becoming the subject of scientific inquiry as well. Science has matured to the point where it can begin to seriously contemplate and even to comprehend life at the level of complexity that we call "landscapes." It is not that we have much choice in this or that this great leap forward is simply an intellectual exercise. It is now abundantly clear that if we do not meet the challenge of landscape ecology, we can harbor little hope of stanching anthropogenic losses in the Earth's biodiversity and hence of stemming the deterioration in the life support system of our own species.

The ontogeny of this writing project began in the autumn of 1991 when the organizers of the Sixth International Theriological Congress to be held in July 1993 in Sydney, Australia, called for symposium suggestions. The time seemed right and the venue appropriate for a serious look at what mammalian ecologists were doing, or perhaps not doing, relative to the rapidly developing field of landscape ecology. In December, I contacted Gray Merriam of Carleton University in Ottawa to seek his support in putting together a symposium proposal. Gray was one of the early leaders in bringing a landscape perspective to the study of mammals; his own published contributions in this area go back to 1979 at least. He agreed, and we assembled an international panel of speakers from among those who have made significant contributions to this area, conceptually or empirically. Including coauthors, we had a group of 12 participants representing nine source countries arranged into eight presentations. Our proposal was accepted, and we thank the Congress organizers, B. J. Fox and M. L. Augee, for their support.

Shortly after, Gray Merriam unfortunately had to withdraw from participation in the Congress. Our plans were too far along at that point to change, and we decided to persevere. Although possible publication of a symposium was in our minds, we decided cautiously to defer discussion of this until after the symposium. When we met for a recap, we decided that we liked what had transpired. In the ensuing months, we explored several possibilities for publication, and we thank John A. Bissonette, Frank B. Golley, Lennart Hansson, and Gray Merriam for their advice and efforts during this critical period.

Discussions with the University of Minnesota Press and their reviewers led us to expand our horizons to include three new chapters in addition to the six chapters based on the original symposium talks (the authors of one presentation were not able to contribute to this volume). In particular, we wanted to expand our treatment of experimental approaches to landscape ecology. Consequently, we invited Gray Merriam to rejoin the project, and he agreed to contribute a chapter (Chap. 4). Gary W. Barrett and Michael S. Gaines also accepted our offer, and we are very pleased to have them and their coauthors as well (Chaps. 8 and 9). The book now contains contributions from 17 authors organized into three parts: context, field approaches, and experimental model systems.

This book has two major themes: the contributions of landscape approaches in mammalian ecology to ecological theory and to conservation biology. In addition, it features a minisymposium on the efficacy of experimental model systems in landscape ecology (Part III). In pursuing these objectives, we anticipate that this book will establish a foundation for what is expected to be a major area of intellectual activity in the future. Both the incorporation of landscape approaches into mainstream ecology and their application to conservation biology will be key elements in this process.

A focus on mammals is important because this group of more than 4,000 species includes many endangered taxa, many large and relatively intelligent forms with complex life histories, and many species that are strong interactors (keystone species) in their respective communities. Thus an understanding of mammals in a landscape context should be a major advance not only toward understanding other species but also toward the conservation of biota in general.

We hope our efforts will draw a diverse audience. For students and others who want an introduction to this new subdiscipline, we offer a conceptual and empirical summary of where the field is today. For landscape ecologists interested in the mammalian perspective, 17 practitioners provide an entrée to their insights and experiences. For wildlife managers and conservation biologists, every chapter relates its subject matter to conservation applications. For ecologists generally, the book breaks new conceptual ground and points to several areas where landscape principles integrate with other subdisciplines. And, for historians of science, three chapters in particular (1, 2, and 8) give different views on whence we came to this point.

Any book is a big project and requires the efforts of many. In addition to those already mentioned, I would like to express collective appreciation to Barbara Coffin and Eileen Griggs, who shepherded the book through review, writing, and preparation for production. None of these steps would have been possible, however, without the dedication, expertise, and hard work of Louise Lidicker, who manipulated and corrected manuscripts in numerous computer languages, formats, e-mail versions, facsimile, and even ordinary paper, not to mention typing correspondence in as many variant modes. Finally, I want to express my gratitude to the 16 other authors and colleagues who did their best for this collective effort in spite of demanding personal schedules and unreasonably short deadlines. For me, it was rewarding!

William Z. Lidicker Jr.
May 1994

Part I

Context: History, Theory, and Conservation

1

The Landscape Concept: Something Old, Something New

William Z. Lidicker Jr.

Roots of Landscape Ecology

Nothing less than a major revolution is beginning to rumble through ecology, and mammalian ecologists will play a major role in this intellectual adventure. These two provocative declarations lie at the heart of this book. I anticipate that the incorporation of landscape concepts and approaches into mainstream ecology will have a beneficial impact on the discipline analogous to that of the incorporation of population and community concepts in earlier decades. A primary beneficiary of these developments will be conservation biology, which carries important corollary implications for land management and related fields.

The landscape notion is not new; it can be traced back to agricultural reforms in late-19th-century Europe (Turner 1989) and to investigations of rodent control in the pre–World War II Soviet Union (Hansson, Chap. 2; Lidicker 1985, 1994b). After World War II, an active discipline began to form in central Europe in an effort to amalgamate ecology, especially phytosociology, with the study of human-modified environments (Naveh 1982; Naveh and Lieberman 1984; Schreiber 1990; Zonneveld 1990). Satellite centers soon developed in North America (Forman and Godron 1986) and Scandinavia (Hansson, Chap. 2). At this early stage, the emphasis was on human-modified landscapes with spatial scales of a kilometer or more, and this remains the perspective of many landscape ecologists (Naveh 1982, 1991; Risser 1987; Risser et al. 1984).

For mammalian ecology, the importance of variation in habitat patch quality was well established in the traditions of wildlife biology (e.g., Leopold 1933), although no attention was given to the specific spatial arrangements of

3

patch types. As early as the 1930s, Soviet ecologists began to emphasize the importance of various combinations of patch types to rodent control. In 1967, P. K. Anderson traveled extensively in the Soviet Union and learned firsthand the views of several leading Soviet ecologists (e.g., B. K. Fenyuk, T. V. Koshkina, N. P. Naumov, P. A. Panteleyev, I. Ya. Polyakov, and S. S. Shvarts). From this experience and his own work on *Mus musculus,* he wrote an important review on ecological structure and gene flow in small mammals (Anderson 1970) in which he proposed that genetic and social fragmentation was indeed the rule for species of small mammals. Then in 1977, Lennart Hansson published his influential paper on the importance of heterogeneous landscapes in the ecology of small mammals. From this time forward, research on landscapes has exploded.

One early manifestation of this new trend has been various attempts to classify habitat patches according to their quality for a focal species; these classifications have often differentiated six grades of quality (Table 1.1). This heuristic pigeonholing of habitat-types has been useful in demolishing the traditional view that species can be found only in their optimal or preferred habitats, with occurrences elsewhere being only temporary, pathological, or accidental, and certainly not of any importance in population dynamics or wildlife management.

In recent years, however, something much more important has been developing with respect to landscape ecology. In 1988, I published the suggestion that landscapes be viewed as a level of organization (biological complexity) above that of communities (Lidicker 1988b). According to this view, landscapes can be defined as ecological systems containing patches of more than one community-type. As such, landscapes are no longer limited to chunks of the biosphere with large spatial dimensions and significant human impacts. Instead they are characterized by a suite of new emergent properties that are not features of the community-types of which they are composed (Table 1.2). Wiens et al. (1993) express a similar view, even listing 11 emergent properties of landscapes. They fail, however, to recognize the holistic nature of their views, and opine instead that "a coherent paradigm for landscape ecology has yet to emerge."

I think one has emerged. In fact, considering landscapes as a new, higher level of organization can be thought of as the latest step in a historical trend beginning with the view common in the early part of this century that ecology is a branch of physiology implying that distributions and abundances are to be understood in terms of physiological adaptations to the abiotic environment (Lidicker 1994b). Two conceptual revolutions followed: first, popula-

Table 1.1. Classification of habitat patches by quality

1	2	3	4	5	References
Constant	Colonization: production	Colonization: reception	Invasion	Traversable	Cockburn 1992
Survival	Colonization	Colonization	Colonization	Traversable	Anderson 1970, 1980
Donor	Induced donor	Reception	Invasion	Transition	Hansson 1977
Primary production	Temporary production	Survival	Invasion	Transition	Myllymäki 1977
Primary	Secondary	Secondary	Secondary	–	Smith et al. 1978
Optimal	Suboptimal	Marginal	Invasive	Traversable	Common usage

Note: Quality decreases from 1 to 5, and category 6 represents completely unsuitable habitat or uncrossable barriers.

tion concepts were incorporated, and then community concepts became integral parts of ecology, with all the familiar benefits ensuing.

Viewing landscapes as a level of biological complexity above that of the community poses a serious semantic problem that must be addressed. The term *landscape* carries the connotation of terrestrial systems, but of course assemblages of community-types can occur in aquatic (waterscapes) as well as terrestrial situations. The same problem afflicts other major languages (German, Spanish, Russian). For this reason, a term like the Finnish *seutu* would be preferable, although it seems unlikely that the world will adopt such a term. (It has, however, accepted another Finnish word, namely, *sauna.*)

Landscapes, Metapopulations, and Metacommunities

Because landscapes, metapopulations, and metacommunities all concern the spatial distribution of biotic arrays, these three concepts may be confused. All three describe discontinuous distributions over space of biotic units at the population or higher level and thus have analogies with population dispersion (spatial distribution of individuals within a population).

Metapopulation

Metapopulation refers to a population of populations, or more precisely, an assemblage of demes of the same species connected by dispersal. The demic

6 WILLIAM Z. LIDICKER JR.

Table 1.2. Some emergent properties of landscapes

Composition (list of community-types)
Diversity (number and relative amounts of community-types)
Spatial configuration (dispersion of patches)
Edge-to-area ratios (sizes and shapes of patches)
Ecotonal features (edge effects)
Connectedness (links among patches of same community-type)
Interpatch fluxes: energy, nutrients, organisms
Dominance relations (among community-types)
Stability: resilience, temporal variations (constancy, predictability), long-term trends (succession, degradation)
Anthropogenic index: measure of human disturbance

structure assures that interdemic dispersal is less than intrademic dispersal. The first formal treatment of metapopulations is often attributed to Levins (1969, 1970), although there were of course numerous antecedents. Levins's classic metapopulation had many unrealistic features but was mathematically tractable. The local populations (demes) were treated like individuals with a given death (extinction) rate and rate of propagule production (emigration). Colonization of empty patches (analogous to births) occurred when an emigrant found such a patch. The demes were all the same size and had no specific spatial arrangement (that is, they were treated as if panmictic), and the interpatch matrix was unspecified, although its hostility could be varied by adjusting the probability of propagule success. The local populations had no internal population dynamics; each deme went from zero to equilibrium numbers immediately following colonization. The rate of change of the metapopulation (in terms of the proportion of patches occupied, p) could then easily be expressed as $dp/dt = mp \, (1 - p) - ep$, where m is the rate of movement of propagules from occupied to unoccupied $(1 - p)$ patches, and e is the extinction rate of patches. Clearly, this is a simple births-minus-deaths growth equation.

Since this auspicious beginning, many more sophisticated and realistic models of metapopulations have been developed (Gilpin and Hanski 1991; Stenseth 1980). It is also clear that metapopulations cover a wide array of situations varying in their degree of demic isolation, reciprocity of dispersal movements, and proneness to global extinction (Harrison 1991). At one extreme are *patchy populations* in which dispersal is frequent enough that extinct local patches are quickly recolonized (rescue effect) (Brown and Kodric-Brown 1977), and the metapopulation is extinction-resistant. At the other extreme is an array of isolated populations (*nonequilibrium metapopulations*) among which dispersal has declined to near zero, making global extinction certain as each local population "winks out" in turn and fails to be recolonized. Somewhere in between are Levins's classic metapopulation and the mainland-island (source-sink) array. The latter is extinction-resistant by virtue of having a single large, stable source population that supplies propagules to less stable satellite patches.

The details of metapopulation dynamics need not concern us here. What is important is the recognition that metapopulations focus on a single species and hence are at the population level of biological complexity. Sometimes when a focal species is viewed in a landscape context, the metapopulation becomes superimposed on a landscape and the two concepts intermingle.

Metacommunity

The much newer idea of *metacommunity* refers to a population of communities, or more precisely an assemblage of patches of the same community-type connected by dispersal. This concept is a natural outgrowth of the metapopulation idea and has many roots in island biogeography (MacArthur and Wilson 1967). Wilson (1992) may have been the first to use the term *metacommunity*. Drake et al. (1993) described the experimental analysis of the assembly of such a metacommunity but mistakenly termed it a landscape assembly.

The important points about metacommunities are three: (1) only a single community-type is involved; (2) the matrix among patches is unspecified and unvarying; and (3) the focus is on the relationships among the patches. In all these respects, metacommunities differ from landscapes, which involve two or more community-types in an array in which the matrix does matter and is clearly specified, and in which all the patches interact meaningfully and in a spatially specific manner. Metacommunities thus represent a community level of biological complexity. It is easy to see, however, how a metacommunity approach can be expanded conceptually into a landscape (*seutu*).

Landscape

By focusing on an ecological system composed of two or more community-types, we achieve a new, higher level of biological complexity. This perspective forces us to consider community patches in a spatially explicit way, and by extrapolation, populations and even individuals can be viewed in a similar manner (Lidicker 1994a). This insight was possibly foreseen by P. Clark and F. Evans, who in 1955 published a paper in *Science* with the following first sentence: "The spatial pattern of distribution of the individual members of a population of organisms is obviously of importance in the analysis of population behavior." It is important also to realize that landscape studies can be pursued on a variety of spatial scales. The scale must be relevant to the question asked and to the focal species or community (Bissonette and Broekhuizen, Chap. 6; Nams 1993; Wiens et al. 1993).

In my view, the landscape approach promises to provide a new and more meaningful level of explanation for ecological phenomena. The examples that I describe below I hope will illustrate the potential of this approach, as will the examples and discussion in all the other chapters of the book. When it is desirable to focus on a single species' metapopulation in a landscape context, the two concepts merge, and we are confronted with the following new and obviously important issues:

1. A focal population may utilize more than one community-type and may even require more than one;

2. Contiguity (juxtaposition) of community-types may have important effects;

3. Distances between patches of the same type and presence or absence of connecting corridors may influence the focal population;

4. Movements of conspecifics in and out of and among habitat patches can affect population dynamics, social behavior, rates of extinction, and genetic composition of populations;

5. Different species may respond differently to a given patch array;

6. Movements of other species (e.g., predators, parasites, prey) in and out of and among habitat patches can be important to a focal species; and

7. Spatial configurations of patches, including areal proportions, edge-to-area ratios, and orientation to interpatch fluxes, can affect population- and community-level processes.

Selected Examples

Several examples of the landscape approach in mammalian ecology will serve to illustrate its potential for significantly improving understanding. They hardly hint at the scope of this effort, or even at what has already been achieved; for this, the entire book is offered as our exemplar. Of all the possible issues facing landscape ecologists, I have chosen only three for this limited exercise.

Patch Quality Differences

The fact that species populations inhabit habitat patches of varying quality is no longer controversial (Table 1.1). Some patches harbor relatively successful source populations (emigration exceeds immigration), and some patches (sinks) contain groups of individuals that are less successful (immigration exceeds emigration) or are successful only some of the time. Arrays of source-sink patches have been modeled by Morris (1991), Ostfeld (1992a), Pulliam (1988), Pulliam and Danielson (1991), and others. Aside from game species, good quantifiable data that detail performance features in specific microhabitats are surprisingly scarce (Morris 1984; Morrison et al. 1992; Ostfeld and Klosterman 1986; Van Horne 1983; Wolff 1980).

One example, the case of the California vole (*Microtus californicus*), will suffice to suggest the ubiquity of this phenomenon. Microhabitats that differ in food quality and cover features have been shown to influence voles in numerous ways. Reproductive success as measured by the number of recruits produced per adult female varies, as do average densities (Cockburn and Lidicker 1983; Krohne 1980; Ostfeld et al. 1985). Recruitment differences translate into source populations for some microhabitats and into sinks for others (Ostfeld and Klosterman 1986; Ostfeld et al. 1985). In the breeding season, adult sex ratios strongly favor females in the better patches; male densities tend to be more homogeneous across microhabitats (Heske 1987; Ostfeld and Klosterman 1986; Ostfeld et al. 1985). The sexes also differ in their response to harsh conditions in the dry season. Persistence in a patch varies seasonally among females, but male persistence is correlated more closely with density (Cockburn and Lidicker 1983; Ostfeld et al. 1985). Finally, extra-large males are found at higher frequency in the marginal microhabitats, where adult females are relatively scarce (Lidicker and Ostfeld 1991).

Juxtaposition of Patch Types

Two examples are provided for this phenomenon, one empirical and one speculative. In a study of cotton rat (*Sigmodon hispidus*) survival through a nonbreeding season (South Carolina), we discovered that adults and subadults of each sex used microhabitats differentially (Lidicker et al. 1992). Dense stands of *Rubus* were used almost exclusively by adult males until the end of the nonbreeding period, when other sex/age groups moved into this patch type. Of particular relevance here is the observation that only *Rubus* patches adjacent to tall grass (*Panicum*) patches were used extensively. We surmise that the *Rubus* provided good cover from predation and the grass-lands were used for foraging. Either patch type alone provided suboptimal habitat.

The second example suggests some speculations on how the density dynamics in a high-quality patch of habitat for a focal species may vary depending on the nature of the habitat patch adjacent to it. In Figure 1.1 the effects of four kinds of neighboring patches are simplified and contrasted. If the favorable patch (A) is surrounded by a large sink (B) (habitat qualities 3–5 in Table 1.1), then dispersal from A will be continuous and unidirectional, so that the focal population density will be strongly influenced by dispersal and will be below carrying capacity in situations with presaturation dispersal. An effective predator living in B could produce this effect even if the sink area were not huge. If patch A is surrounded by a barrier habitat (C) (quality 6 in Table 1.1), density fluctuations may show a strong seasonal pattern. Alternatively, the population in A may increase so as to exceed and destroy its carrying capacity and then crash to extinction or a precariously low number. This outcome would be especially likely if patch A were small and/or the focal species could destroy its food supply (herbivores that are folivores or browsers, specialized carnivores or parasites).

The last two scenarios involve adjacent patches that are refuges for predators of the focal species. If the predator is specialized to feed on the target species (patch D, Fig. 1.1c), several outcomes are possible. If the predator remained in its refuge patch, only dispersers would be available to it, and the patch would act like a habitat sink (patch B). More typically, however, the specialized predator would invade patch A and respond numerically to increases in prey density. The result could be that prey numbers would decline to low levels, which would be followed by a decline in the predator, and multiannual cycles would be generated. On the other hand, if the predator were generalized and the matrix were not a large sink (patch E, Fig. 1.1d), then the predator's response to changes in prey density would be mainly functional rather than numerical, as alternative prey would presumably be available. Predator numbers would not decline when prey declined. Consequently, prey densities would show less intense annual cycles at moderate levels. (For further discussion on the role of predator refuges in the context of empirical data, see Oksanen and Schneider [Chap. 7] and Oksanen 1993.)

Proportions of Different Habitat-Types

Organisms clearly are sensitive to the spatial array of their optimal habitat patches in a landscape, if for no other reason than because the distances be-

Patch contiguity

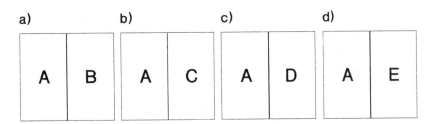

Density dynamics

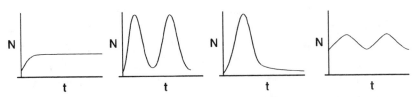

Fig. 1.1. A speculative view of how the nature of patch contiguity can influence density dynamics in a focal species. Patch A is designated as an optimal habitat patch for the focal species of interest. Patch B is a large sink habitat; C is a barrier habitat (dispersal is prevented); D is a refuge habitat for a predator specialized to feed on the target species; and E is a refuge habitat for a generalized predator. N is the size of the focal population, and t is time.

tween patches and the presence or absence of dispersal corridors are critical to the persistence of the resulting metapopulation (Kozakiewicz and Szacki, Chap. 5; Laurance, Chap. 3; Merriam, Chap. 4; van Apeldoorn et al. 1992a,b). The nature of the interpatch matrix will interact with distance to influence the rate of successful dispersal among patches (Lankester et al. 1991; Verboom and van Apeldoorn 1990; Zhang and Usher 1991).

One aspect of these landscape influences is their impact on the population dynamics of organisms within patches (see also previous section). A preliminary attempt to conceptualize this with respect to microtine rodent multiannual cycles was the ROMPA hypothesis (Hansson, Chap. 2; Lidicker 1985, 1988a, 1991; Ostfeld 1992b). ROMPA refers to the ratio of optimal to marginal patch areas and hence expresses the proportion of a landscape composed of optimal habitat for the target species. The idea was that this ratio could influence the probability that a population would undergo multiannual cycles in numbers. If optimal habitat were scarce (low ROMPA), then the matrix would serve as a large sink, and densities would be fairly stable (Fig. 1.1a). At very high ratios, densities would also be stable or show annual cycles, but at intermediate ROMPA, multiannual cycles would be likely. This is because there would be enough suboptimal or marginal habitat that it would not fill to capacity in one breeding season or even two (possibly aided by specialist predators). Once it was filled, however, rapidly increasing densities and frustrated dispersal would follow, leading to density crashes. Ostfeld (1992b) has built on this idea by suggesting that the quality of the matrix is clearly important in determining the demographic result and that in general the greater the difference in quality between patch and matrix, the more likely it is that densities will be stable.

Delattre et al. (1992) provide empirical data on *Microtus arvalis* that support the importance of patch proportions to population dynamics. They compiled data from eight sites across northern and central France and used a landscape perspective to examine demographic factors in this economically important species of vole. They describe five types of vole demographic patterns, based on three criteria: annual vs. multiannual fluctuations, the amplitude of density changes, and the presence or absence of local extinctions. At five sites, they also assessed populations of a major vole predator, the weasel *Mustela nivalis*.

In general, the tendency of voles to show high-intensity fluctuations and multiannual cycles increased as the percentage of the landscape devoted to fallow cropland and permanent grass increased. This tendency was reflected in the fact that the percentage of municipalities reporting vole damage to crops was positively but nonlinearly related to the percentage of land in grass cover (Fig. 1.2a). When the percentage of grass cover exceeded 50%, multiannual cycles tended to change to strong annual fluctuations. Similarly, municipalities with higher percentages of forest cover suffered less vole damage (Fig. 1.2b). The authors suggest that this stabilizing effect of forest is caused by five species of generalist predators that live primarily in the forest but forage outside as well (Fig. 1.1d). On the other hand, numbers and amplitudes of the weasel, a specialist predator, increased in proportion to vole densities and were positively associated with multiannual cycles (Fig. 1.1c).

The results of Delattre et al. (1992) support the ROMPA hypothesis since numbers were low and relatively steady at low values of ROMPA (< 5% grassland), exhibited multiannual cycles at medium values (5–50%), and exhibited annual cycles only at high ROMPA (> 50%). The ROMPA hypothesis also apa)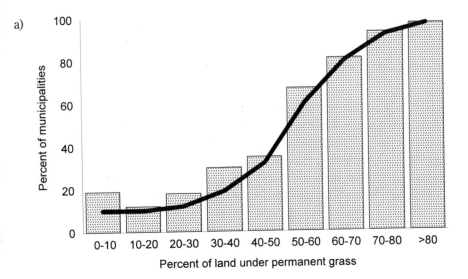

b)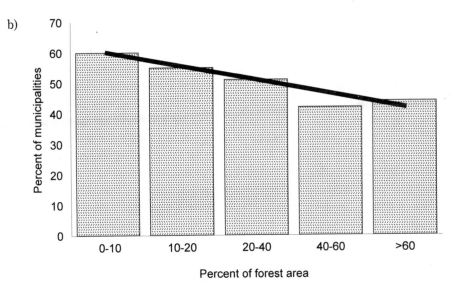

Fig. 1.2. Percentage of municipalities in France reporting crop damage by *Microtus arvalis* (a) as a function of the percentage of the agricultural areas under permanent grass cover and (b) as a function of the percentage of the municipality in forest. Redrawn from *Agriculture, Ecosystems and Environment*, vol. 39, Delattre et al., "Land use patterns and types of common vole *(Microtus arvalis)* population kinetics," pp. 153–68, copyright 1992, with kind permission from Elsevier Science Ltd., The Boulevard, Langford Lane, Kidlington OX5 1GB, U.K.

pears to be a special case of the modified trophic exploitation hypothesis discussed by Oksanen and Schneider (Chap. 7).

Conservation Biology

The re-emergence of conservation biology in recent decades as a respectable scientific discipline is inextricably intertwined with the development of landscape ecology. Conservation is an inherently holistic notion in that attention must be directed to the behavior, physiology, and genetics of individuals, to population dynamics, to community processes, and most especially to landscape phenomena, not to mention social, political, economic, and policy complications. Until holistic approaches could be incorporated into mainstream ecology, conservation was necessarily viewed as primarily a political and aesthetic movement (see Lidicker 1994b for a more thorough treatment of this history). In the 1980s, science began to catch up with conservation and, along with the accelerating deterioration of the human life support system, conservation became established with a firm foundation in basic science (Cox 1993; Fiedler and Jain 1992; Soulé 1986).

Conservation biology is directed toward reducing losses in the Earth's biodiversity. Biodiversity has four aspects: (1) species richness, (2) genetic variation within populations, (3) genetic variation across the distribution of species, reflecting local adaptation and the raw materials for future evolutionary change, and (4) the complex of ecological interactions (coactions) in which organisms are involved. It is not sufficient to save species in isolation (i.e., captive breeding); rather, species must be preserved as functioning members of their communities and on a large enough scale that the communities have long-term stability. The rationale for this focus on conservation of biodiversity is ultimately the assumption that humans desire to survive and that preferably their future should be civilized and above a subsistence standard of living. Therefore, as losses to our planet's biodiversity render our life support system increasingly precarious, conservation biologists are contributing their efforts to improving human welfare in both the present and the future.

The preservation of biodiversity is opposed by increasing fragmentation of community-types through human activities, coupled with assaults from environmental pollution and the incessant transport of species, both intentionally and accidentally, around the world. Extinction rates are difficult to estimate because only 6–20% of the estimated total of the Earth's biota has been scientifically inventoried. Depending on one's intuition of how many undescribed species there are, current extinction rates can be estimated at one to four orders of magnitude higher than background rates since the end of the Cretaceous (Primack 1993; Raup 1978; Wilson 1988). The challenge for conservation is clearly immense.

Metapopulations, metacommunities, and landscapes are all concepts that concern the behavior of organisms in a fragmented environment. This is not a new situation for organisms, but it is certainly a much more common circumstance than existed before the last few hundred years. Thus landscape ecology would have been a beneficial development for biology even without the recent history of anthropogenic fragmentation, but given that history it is an essential and overdue development. We now have a conceptual framework in which to analyze the influence of habitat patch size and shape, the role of juxtaposition of different kinds of communities, the importance of corridors among patches, the influence of edge effects, and the impact of varying proportions and qualities of different community-types in a landscape. A recent effort to address a major conservation issue from a landscape perspective is that on behalf of the Northern Spotted Owl (*Strix occidentalis*) in western North America (Thomas et al. 1990). A comparable European effort is that for the badger (*Meles meles*) in The Netherlands (Lankester et al. 1991).

Conservation efforts would be difficult enough even if all the problems were strictly scientific. In reality, strong cultural, political, and economic interests oppose these efforts, ironically from the very constituency that conservation biologists seek to help. Pessimism, however, is self-fulfilling.

Summary

This chapter presents a brief history of landscape concepts culminating in a postulated new threshold of understanding for mammalian ecology. Landscapes can be defined as ecological systems containing patches of more than one community-type. Thus, they are characterized by a number of emergent properties that are not features of their constituent community-types. This holistic view places landscapes at a level of biological organization above that of the community. As such, landscapes occur in both terrestrial and aquatic situations and can vary greatly in absolute size. In contrast, metapopulations are assemblages of demes (same species) connected by dispersal and hence represent the population level of complexity. And, metacommunities are assemblages of patches of the same community-type connected by dispersal and thus represent the community level of organization. These new developments can be viewed as the latest step in a historical trend that began with the view that ecology is a branch of physiology implying that distributions and abundances are to be understood in terms of physiological adaptations to the abiotic environment. Population- and community-level perspectives followed in turn with familiar benefits.

Selected examples illustrate just three landscape issues: (1) the demographic and social effects of patch quality differences, (2) the relevance of patch-type juxtapositions, and (3) the influence on demography of different proportions of community-types in the landscape.

A corollary of the new developments in landscape ecology is the revitalization of conservation biology, which is directed toward reducing losses to the Earth's biodiversity. Concepts concerning the behavior of organisms in fragmented environments provide a framework for analyzing a variety of conservation issues.

Acknowledgments

My thanks go to Lauri and Tarja Oksanen for providing information on the Finnish word *seutu*. L. Hansson made helpful comments on the first draft. L. N. Lidicker contributed in many ways, including tremendous assistance in preparation of the figures and manuscript.

Literature Cited

Anderson, P. K. 1970. Ecological structure and gene flow in small mammals. *Symp. Zool. Soc. London* 26:299–325.
Anderson, P. K. 1980. Evolutionary implications of microtine behavioral systems on the ecological stage. *The Biologist* 62:70–88.
Brown, J. H., and A. Kodric-Brown. 1977. Turnover rates in insular biogeography: Effect of immigration on extinction. *Ecology* 58:445–9.
Clark, P. J., and F. C. Evans. 1955. On some aspects of spatial pattern in biological populations. *Science* 121:397–8.
Cockburn, A. 1992. Habitat heterogeneity and dispersal: Environmental and genetic patchiness. Pages 65–95 *in* Stenseth, N. C., and W. Z. Lidicker Jr. (eds.), *Animal Dispersal: Small Mammals as a Model*. London: Chapman & Hall. 365 pp.
Cockburn, A., and W. Z. Lidicker Jr. 1983. Microhabitat heterogeneity and population ecology of an herbivorous rodent, *Microtus californicus. Oecologia* 59:167–77.
Cox, G. W. 1993. *Conservation Ecology.* Dubuque, Iowa: Wm. C. Brown. 352 pp.
Delattre, P., P. Giraudoux, J. Baudry, et al. 1992. Land use patterns and types of common vole (*Microtus arvalis*) population kinetics. *Agric. Ecosyst. Environ.* 39:153–68.
Drake, J. A., T. E. Flum, G. J. Witteman, et al. 1993. The construction and assembly of an ecological landscape. *J. Anim. Ecol.* 62:117–30.
Fiedler, P. L., and S. K. Jain (eds.). 1992. *Conservation Biology: The Theory and Practice of Nature Conservation, Preservation, and Management.* New York: Chapman & Hall. 507 pp.
Forman, R. T. T., and M. Godron. 1986. *Landscape Ecology.* New York: John Wiley & Sons. 620 pp.
Gilpin, M., and I. Hanski (eds.). 1991. Metapopulation dynamics: Empirical and theoretical investigations. *Biol. J. Linn. Soc.* 42:1–336.

Hansson, L. 1977. Spatial dynamics of field voles *Microtus agrestis* in heterogeneous landscapes. *Oikos* 29:539–44.

Harrison, S. 1991. Local extinction in a metapopulation context: An empirical evaluation. *Biol. J. Linn. Soc.* 42:73–88.

Heske, E. J. 1987. Spatial structuring and dispersal in a high density population of the California vole, *Microtus californicus. Holarctic Ecol.* 10:137–49.

Krohne, D. T. 1980. Interspecific litter size variation in *Microtus californicus.* II. Variation between populations. *Evolution* 34:1174–82.

Lankester, K., R. van Apeldoorn, E. Meelis, and J. Verboom. 1991. Management perspectives for populations of the European badger (*Meles meles*) in a fragmented landscape. *J. Appl. Ecol.* 28:561–73.

Leopold, A. 1933. *Game Management.* New York: Charles Scribner. 481 pp.

Levins, R. A. 1969. Some demographic and genetic consequences of environmental heterogeneity for biological control. *Bull. Entomol. Soc. Am.* 15:237–40.

Levins, R. A. 1970. Extinction. *Lect. Math. Life Sci.* 2:75–107.

Lidicker, W. Z., Jr. 1985. Population structuring as a factor in understanding microtine cycles. *Acta Zool. Fennica* 173:23–7.

Lidicker, W. Z., Jr. 1988a. Solving the enigma of microtine "cycles." *J. Mammal.* 69:225–35.

Lidicker, W. Z., Jr. 1988b. The synergistic effects of reductionist and holistic approaches in animal ecology. *Oikos* 53:278–81.

Lidicker, W. Z., Jr. 1991. In defense of a multifactor perspective in population ecology. *J. Mammal.* 72:631–5.

Lidicker, W. Z., Jr. 1994a. A spatially explicit approach to vole population dynamics. *Pol. Ecol. Stud.* 20:215–25.

Lidicker, W. Z., Jr. 1994b. Population ecology. Pages 322–46 *in* Birney, E. C., and J. R. Choate (eds.), *Seventy-five Years of Mammalogy (1919–1994).* Spec. Publ. 11. American Society of Mammalogists. 433 pp.

Lidicker, W. Z., Jr., and R. S. Ostfeld. 1991. Extra-large body size in California voles: Causes and fitness consequences. *Oikos* 61:108–21.

Lidicker, W. Z., Jr., J. O. Wolff, L. N. Lidicker, and M. H. Smith. 1992. Utilization of a habitat mosaic by cotton rats during a population decline. *Landscape Ecol.* 6:259–68.

MacArthur, R. H., and E. O. Wilson. 1967. *The Theory of Island Biogeography.* Princeton, N.J.: Princeton University Press. 203 pp.

Morris, D. W. 1984. Sexual differences in habitat use by small mammals: Evolutionary strategy or reproductive constraint? *Oecologia* 65:51–7.

Morris, D. W. 1991. On the evolutionary stability of dispersal to sink habitat. *Am. Nat.* 137:907–11.

Morrison, M. L., B. G. Marcot, and R. W. Mannan. 1992. *Wildlife Habitat Relationships: Concepts and Applications.* Madison: University of Wisconsin Press. 343 pp.

Myllymäki, A. 1977. Demographic mechanisms in the fluctuating populations of the field vole *Microtus agrestis. Oikos* 29:468–93.

Nams, V. O. 1993. Spatial scale of habitat selection. Pages 220–1 *in* Augee, M. L. (ed.), *Abstracts Sixth International Theriological Congress.* Sydney: Congress Organizing Committee. 344 pp.

Naveh, Z. 1982. Landscape ecology as an emerging branch of human ecosystem science. *Adv. Ecol. Res.* 12:189–237.

Naveh, Z. 1991. Some remarks on recent developments in landscape ecology as a transdisciplinary ecological and geographical science. *Landscape Ecol.* 5:65–73.

Naveh, Z., and A. S. Lieberman. 1984. *Landscape Ecology: Theory and Application.* New York: Springer-Verlag. 365 pp.

Oksanen, T. 1993. Does predation prevent Norwegian lemmings from establishing permanent populations in lowland forests? Pages 425–37 *in* Stenseth, N. C., and R. A. Ims (eds.), *The Biology of Lemmings*. London: Academic Press. 683 pp.

Ostfeld, R. S. 1992a. Effect of habitat patchiness on population dynamics: A modelling approach. Pages 851–63 *in* McCullough, D. R., and R. H. Barrett (eds.), *Wildlife 2001: Populations*. London: Elsevier Applied Science. 1,163 pp.

Ostfeld, R. S. 1992b. Small-mammal herbivores in a patchy environment: Individual strategies and population responses. Pages 43–74 *in* Hunter, M. D., T. Ohgushi, and P. W. Price (eds.), *Effects of Resource Distribution on Animal-Plant Interactions*. London: Academic Press. 505 pp.

Ostfeld, R. S., and L. L. Klosterman. 1986. Demographic substructure in a California vole population inhabiting a patchy environment. *J. Mammal.* 67:693–704.

Ostfeld. R. S., W. Z. Lidicker Jr., and E. J. Heske. 1985. The relationship between habitat heterogeneity, space use, and demography in a population of California voles. *Oikos* 45:433–42.

Primack, R. B. 1993. *Essentials of Conservation Biology.* Sunderland, Mass.: Sinauer Associates. 564 pp.

Pulliam, H. R. 1988. Sources, sinks and population regulation. *Am. Nat.* 132:652–61.

Pulliam, H. R., and B. J. Danielson. 1991. Sources, sinks, and habitat selection: A landscape perspective on population dynamics. *Am. Nat. (Suppl.)* 137:S50–66.

Raup, D. M. 1978. Cohort analysis of generic survivorship. *Paleobiology* 4:1–15.

Risser, P. G. 1987. Landscape ecology: State-of-the-art. Pages 3–14 *in* Turner, M. G. (ed.), *Landscape Heterogeneity and Disturbance*. New York: Springer-Verlag. 239 pp.

Risser, P. G., J. R. Karr, and R. T. T. Forman. 1984. *Landscape Ecology, Directions and Approaches.* Spec. Publ. 2. Champaign: Illinois Natural History Survey. 18 pp.

Schreiber, K. F. 1990. The history of landscape ecology in Europe. Pages 21–33 *in* Zonneveld, I. S., and R. T. T. Forman (eds.), *Changing Landscapes: An Ecological Perspective*. New York: Springer-Verlag. 286 pp.

Smith, M. H., M. N. Manlove, and J. Joule. 1978. Spatial and temporal dynamics of the genetic organization of small mammal populations. Pages 99–113 *in* Snyder, D. P. (ed.), *Populations of Small Mammals under Natural Conditions*. Pymatuning Symposia in Ecology, Spec. Publ. Ser., Vol. 5. Pittsburgh: Pymatuning Laboratory of Ecology. 237 pp.

Soulé, M. E. (ed.). 1986. *Conservation Biology: The Science of Scarcity and Diversity.* Sunderland, Mass.: Sinauer Associates. 584 pp.

Stenseth, N. C. 1980. Spatial heterogeneity and population stability: Some evolutionary consequences. *Oikos* 35:165–84.

Thomas, J. W., E. D. Forsman, J. B. Lint, E. C. Meslow, B. R. Noon, and J. Verner. 1990. *A Conservation Strategy for the Northern Spotted Owl.* Portland, Oreg.: USDA Forest Service, USDA Bureau of Land Management, USDI Fish and Wildlife Service, and USDI National Park Service. 427 pp.

Turner, M. G. 1989. Landscape ecology: The effect of pattern on process. *Annu. Rev. Ecol. Syst.* 20:171–97.

van Apeldoorn, R., H. Hollander, W. Nieuwenhuizen, and F. van der Vliet. 1992a. De Noordse woelmuis in het deltagebied. *Landschap* 9:189–202.

van Apeldoorn, R. C., W. T. Oostenbrink, A. van Winden, and F. F. van der Zee. 1992b. Effects of habitat fragmentation on the bank vole, *Clethrionomys glareolus,* in an agricultural landscape. *Oikos* 65:265–74.

Van Horne, B. 1983. Density as a misleading indicator of habitat quality. *J. Wildl. Manage.* 47:893–901.

Verboom, B., and R. van Apeldoorn. 1990. Effects of habitat fragmentation on the red squirrel, *Sciurus vulgaris* L. *Landscape Ecol.* 4:171–6.

Wiens, J. A., N. C. Stenseth, B. Van Horne, and R. A. Ims. 1993. Ecological mechanisms and landscape ecology. *Oikos* 66:369–80.

Wilson, D. S. 1992. Complex interactions in metacommunities, with implications for biodiversity and higher levels of selection. *Ecology* 73:1984–2000.

Wilson, E. O. 1988. The current state of biological diversity. Pages 3–18 *in* Wilson, E. O. (ed.), *Biodiversity.* Washington, D.C.: National Academy Press. 521 pp.

Wolff, J. O. 1980. The role of habitat patchiness in the population dynamics of snowshoe hares. *Ecol. Monogr.* 50:111–30.

Zhang, Z., and M. B. Usher. 1991. Dispersal of wood mice and bank voles in an agricultural landscape. *Acta Theriol.* 36:239–45.

Zonneveld, I. S. 1990. Scope and concepts of landscape ecology as an emerging science. Pages 3–20 *in* Zonneveld, I. S., and R. T. T. Forman (eds.), *Changing Landscapes: An Ecological Perspective.* New York: Springer-Verlag. 286 pp.

2

Development and Application of Landscape Approaches in Mammalian Ecology

Lennart Hansson

Early Landscape Approaches

Interest in the importance of landscape composition (or of the effects of the interspersion of optimal, marginal, and uninhabitable habitats) for the distribution and abundance of mammals probably developed first in the Soviet Union between the two world wars. Many steppe environments harbor endemic diseases transmitted by rodents, especially gerbils. Applied studies on the diseases and vectors led to the finding that stable foci of disease can occur only in specific landscapes, either along river valleys or in islandlike patches in the steppes (Naumov 1936, 1948, 1972, 321–3). Permanent burrows in such locations are abandoned during temporary but local disease outbreaks. However, they are soon resettled because they are located in dispersal corridors or because of local improvements in food supply achieved by rodent activity (fertilization of the close neighborhood of a colony by middens and feces). Favorable habitat patches were thus easily located by dispersing animals. In more homogeneous and extensive habitats, in contrast, epizootic diseases wipe out entire populations of rodents as soon as they start to become abundant again.

The ideas of Naumov (1936) and his co-workers were discussed in one early paper based on research in Great Britain by Evans (1942), who, however, did not accept the concept of "favorable habitats" that were occupied more or less continuously. Evans argued instead that dense local populations are easily destroyed by predation or disease. He suggested that "habitats which will maintain only low densities may be essential to the ultimate survival of a species."

The appreciation of the significance of landscape composition for rodent dynamics was apparently transferred to North America by P. K. Anderson (1970, 1980, 1989), who had learned about the experience of the Soviet ecologists, evidently during visits to the Soviet Union. This experience influenced his ideas about permanent and temporary habitats (Naumov 1948, 1972, 323), especially of the house mouse, *Mus musculus* (Anderson 1970). Other early work in North America included that of Stickel (1979), who demonstrated the use of alternative habitats by house mice in temporary cropland, and Wegner and Merriam (1979), who stressed the importance of wooded fencerows for the use of whole landscapes (e.g., agricultural areas and homesteads) by woodland rodents and birds. Such problems were considered to belong to one of the main approaches to landscape ecology at a landmark meeting of North American landscape ecologists (Risser et al. 1984).

On the other hand, work on the effects of patch distribution and corridors in Canada (Merriam 1984) was considered to be more related to den Boer's (1968, 1981) concept of "spreading of risk." Den Boer's ideas of regional population persistence in certain patches under catastrophic impact on other patches and of response variations due to differing demographic composition in various patches were obviously influential. Work on parasites and diseases by Per Brinck at the University of Lund in Sweden in collaboration with a Slovak parasitologist, Milan Mrciak, well acquainted with Naumov's work (Edler and Mrciak 1975) made the eastern ideas known in Scandinavia in the early 1970s and may have affected ecological thinking there.

Thus, applied research first demonstrated landscape effects; more basic or theoretical approaches do not seem to occur in the readily available mammalogical literature until the 1970s (Anderson 1970; Hansson 1977; Stenseth 1980; Wegner and Merriam 1979). These new ideas were primarily related to stability and variability in rodent populations. At the same time the importance of landscape heterogeneity for pest control, and especially for prevention of forest damage by small rodents, was stressed (Stenseth 1977; Stenseth and Hansson 1981). In the early 1980s, work began also on the effects of landscape composition on the persistence of local populations. This work had both a theoretical approach (Lefkovitch and Fahrig 1985) and an applied one (Henderson et al. 1985), especially related to conservation of small and endangered populations. The applied approach emphasized landscape configurations that permitted population persistence.

Island biogeography and metapopulation dynamics contain elements of environmental heterogeneity and landscape ecology, but they derive from the early-recognized ecological subdisciplines of biogeography (MacArthur and Wilson 1967) and pure pest control (Levins 1969), respectively. Island biogeographic theory of mammal communities developed independently of landscape ecology in the early 1980s (Lomolino 1982, 1986; Western and Ssemakula 1981). The equilibrium biogeographic theory was even rejected or considered equivocal for most continental patchy situations by several landscape ecologists working with mammals (Merriam and Wegner 1992; Middleton and Merriam 1983; Risser et al. 1984). First, island biogeography considered species numbers and not the separate species populations and their extinctions and recolonizations. Second, the heterogeneous terrestrial environment around habitat islands was easily observed to be inconsistent with the homogeneous matrix assumed in the equilibrium island dynamics. Furthermore, island biogeographic theory assumes comparatively limited immigration rates and cannot explain causal relationships regarding variations in movements to habitat islands. Metapopulation theory has been better accepted by landscape ecologists, although many of the original assumptions have been rejected; as a result, *metapopulation* can at present refer both to equilibrium dynamics between subpopulation extinctions and recolonizations in habitat patches (Gilpin and Hanski 1991) and to any subdivided population (Merriam et al. 1989).

Figure 2.1 depicts an array of landscape structures considered in the early work. The development of landscape approaches in mammalogy can easily be divided into effects on abundance and dynamics on the one hand and effects on population persistence on the other. Both approaches can then be divided into theoretical and applied aspects.

Landscape Composition and Population Dynamics

Theoretical Approaches

Anderson (1970) proposed that *Mus* populations could fairly strictly be divided into permanent and temporary ones; the latter usually succumbed in autumn or winter, and there was little or no return migration. He later developed this approach in relation to the idea of dispersal as nonadaptive behavior (Anderson 1989).

Hansson (1977) stressed the importance of landscape heterogeneity—especially the occurrence of habitats with low survival rates—for the stability of vole populations. The emphasis was on environments with more or less shelter against predators for folivorous rodents. The importance of dispersal, and especially failed dispersal, was stressed. Similarly, Stenseth (1980) appears also to have had mainly predation in mind in his more theoretical analyses. However, he may have been the first to propose that landscape composi-

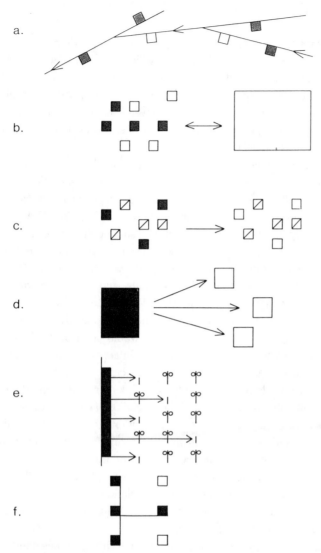

Fig. 2.1. Some of the original ideas in landscape ecology; shaded blocks represent occupied patches, and unshaded ones are empty. (a) Colonies in dispersal corridors are generally occupied but may temporarily go extinct (e.g., due to disease). (b) Patchily distributed populations show a higher level of persistence than continuous populations during pest outbreaks. (c) Patchily distributed high-density populations (black blocks) are more exposed to, for example, predation and disease than are low-density populations (blocks with oblique lines). (d) Dispersal from permanent populations only leads to temporary establishment of daughter populations. (e) Refugial populations may often enter or invade neighboring habitats, for example, those with sensitive crops (dicot symbols). (f) Isolated patches do not support persistent subpopulations; connectivity by various means is important.

tion determines whether a rodent population, independently of nutritive or social adaptations, will be cyclic or noncyclic (Stenseth 1980). I believe that this proposition has caused many problems in subsequent analyses. Furthermore, a more general paper on dispersal by Lidicker (1975), suggesting both adaptive and nonadaptive movements between habitat patches, although it did not treat landscape ecology per se, has been very influential in later discussions of landscape ecology.

To put the discussion of the importance of landscape heterogeneity for population dynamics on a more concrete basis, Lidicker (1988) introduced the concept of ROMPA, the ratio of optimal to marginal patch area (Fig. 2.2). Lidicker (1988) suggested that landscapes with high ROMPA would not demonstrate cycles because there would be so much good habitat during severe seasons that sufficient numbers would remain to enable complete recovery during the following breeding season. At low ROMPA, only a few animals would survive in the limited optimal habitats, and the population recovery would take several years. Thus, low ROMPA would often cause cycles.

Gaines et al. (1991) argued in what seemed to be exactly the opposite way. At high ROMPA, the few dispersal sinks would soon be filled, and there should be frustrated dispersal in optimal habitats. The ensuing high densities might exceed the carrying capacity and crash. In landscapes with low ROMPA, there would instead be extensive sinks for dispersing animals, and populations in optimal habitats should be stabilized at intermediate densities. However, there seems to have been a misunderstanding between Gaines et al. (1991) and Lidicker (1988), in that the former's "high" and the latter's "low" both refer to medium-range densities, and thus there does not appear to be any significant disagreement (Lidicker 1991, Chap. 1; Ostfeld 1992a).

Ostfeld (1992a) again reanalyzed these relationships and introduced the relative quality of habitat patches. If the optimal habitat is much better than the suboptimal (or rather marginal) habitats, then individuals dispersing from the good patches will likely succumb, and the dynamics should generally be stable in optimal patches. However, if ROMPA is relatively large, there might be a potential for irregular outbreaks, as even the poor habitat would fill up. On the other hand, if there are only small quality differences between optimal and suboptimal habitats, then high ROMPA is supposed to lead to only intra-annual (seasonal) variation, and low ROMPA to multiannual cycles. Ostfeld (1992a) also stressed the importance of seasonal as well as relative variations in patch quality. However, Ostfeld's (1992a) ideas can be seen partly as a modification and extension of Lidicker's (1988) more original hypothesis.

There are few empirical tests of these ideas. Foster and Gaines (1991) examined landscape effects in experimental landscapes with constant quality in optimal and suboptimal grassland areas (see also Diffendorfer et al., Chap. 9). The size of the optimal patches affected species distribution: the largest species were most abundant in large patches, and small species were most abundant in smaller patches. The matrix, assumed to be a marginal or uninhabitable habitat, supported large numbers of the granivorous *Peromyscus maniculatus* but few other rodents. Thus, species differed in the use of large and small patches and in the use of areas assumed to be optimal and marginal. The dynamics were stable in all patches, as far as could be judged from the three-year study.

Ostfeld (1992a) considered the dynamics of two *Microtus californicus* populations in relation to habitat patchiness. Both populations lived in habitats with low ROMPA, but the suboptimal habitats of one (Ostfeld and Klosterman 1986) were much worse than those of the other (Ostfeld et al. 1985). The former population was stable, while the latter was cyclic. The quality of the various habitats in all these studies was assessed with regard to the nutritional value of the main vegetation and not its protection value against predation.

Ostfeld (1992b) performed simulation analyses of rodent populations in landscapes with optimal and marginal patches, with varying sizes of marginal patches, resource levels, and social systems. The dynamics of an originally cyclic population in the optimal patch were in fact not particularly affected except with a fairly unrealistic social system, that is, strict female territoriality. The model may apply to the partly granivorous *Clethrionomys glareolus* (Bujalska 1990). However, the greatest stabilizing effect with more realistic semiterritorial females was in a situation with marginal patches with fairly high food quality but also a high predation rate (Ostfeld 1992b, 860).

If we compare the general ideas in the ROMPA hypothesis (unspecific rodents in habitats of varying quality) with the ideas from the 1970s (Hansson 1977; Stenseth 1980), then it is clear that different structures and mechanisms have been proposed for the limitation of populations. In the early hypotheses, transition areas were included as especially hostile environments that often had to be traversed (Fig. 2.2). The limiting mechanisms were then assumed to act at least partly in species-specific ways. Mortality due to predation or starvation could cause different dynamics in various species. An animal of whatever kind taken by a predator while moving into transition areas or inside an exposed suboptimal patch would immediately disappear, while food limitation (implicitly assumed to be an eventual limiting mechanism in later papers) would act differently depending on the nutritive adaptations of the species concerned. Carnivores such as shrews and granivorous mice might die promptly with a food deficit, whereas folivores (voles) and bryovores (lemmings) might be affected more gradually because they can survive on low-quality food. As food levels decrease, the latter species may instead decline in body size, reproduction, immunity against parasites and disease, and possibly antipredator behavior. The effects on granivores and folivores will be similar when populations are limited by predation in marginal patches; in food-limited patches, granivores will soon die, whereas folivores may continue to increase in numbers while declining in individual quality. Finally, dense, evenly distributed populations of folivores (Hansson 1990) in poor physiological shape will crash from starvation, disease, or attraction of specialist predators: cyclic dynamics will primarily appear in the folivores (Hansson 1994).

Applied Aspects

The first landscape approaches in pest control were related to the importance of refugia in providing propagules that invade and colonize economically sensitive habitats. Such refugia could be abandoned fields or permanent grasslands for voles attacking alfalfa and other important agricultural crops in Germany (Frank 1956) or reedbeds for *Mus* populations invading wheat fields in Australia (Newsome 1969a,b). This knowledge was evidently also put to use in practical pest control measures; for example, the plowing of grasslands or removal of old grass close to forest plantations and seed orchards to discourage invasions by voles in winter under the snow (Larsson 1975).

A practical pest control strategy based on landscape composition was derived by Stenseth and Hansson (1981), who analyzed metapopulation models following Levins (1969) and Stenseth (1981). They concluded that if there are empty patches for pest rodents and if there are inexpensive and effective methods of reducing immigration, then most resources should be put into such preventive measures, but in more realistic pest situations with dense populations and mainly density-independent limitation, extinction should be sought via control treatments distributed as evenly as possible in time and as randomly as possible in space.

Redhead et al. (1985), working in semiarid Australia, seem to have been the first to consider both the practical importance of refugia and the importance of including refugia in a theoretical framework of landscape functioning. They identified donor habitats (i.e., refugia), transition areas (arid fields during most years), and induced donor habitats (e.g., wheatfields after extensive rains), as defined by Hansson (1977, Table 1.1). They stressed that con-

Fig. 2.2. Two ways of looking at landscape heterogeneity as a potential stabilizing factor in rodent dynamics. The top two diagrams illustrate ideas from the 1970s, with optimal areas (black); suboptimal habitats (white), normally with negative population growth but sometimes transformed by heavy immigration into induced donor habitats; and transition areas (hatched), uninhabitable except for dispersal. Landscape composition was thought to affect short-term dynamics, particularly of folivorous species. The two bottom diagrams depict ideas from the 1980s and 1990s: ratios of optimal (black) to marginal (white) patch areas (ROMPAs) are assumed to affect total numbers and long-term dynamics in a landscape. The marginal area is supposed to temporarily support the animals, while uninhabitable areas are not considered. Considering that transition areas are necessary, I have here assumed that marginal areas are always in excess.

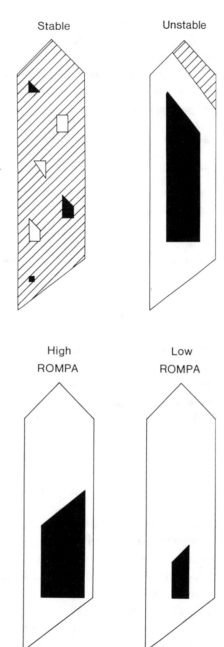

trol, either by poisoning or by habitat modification, should be directed primarily toward donor habitats. The identification of landscape-dependent dynamics has led to the formulation of a comprehensive control strategy known as PICA (predict, inform, control, and assess) (Redhead 1988). Later work has supported the importance of induced donor habitats in leading to extensive pest outbreaks in certain areas (Singleton 1989).

Landscape Composition and Population Persistence

Persistence of Local Populations and Metapopulations

Smith (1974) probably first hinted at the importance of landscape composition for the persistence of outlying small populations in studies of pikas (*Ochotona princeps*) living in a mosaic of talus slopes. These studies were only much later put into a metapopulation perspective (Ray et al. 1991).

The persistence of (meta)populations in landscapes consisting of several or many subpopulations separated by unsuitable land has been treated in a comprehensive way, both theoretically (Lefkovitch and Fahrig 1985) and in field work on rodents (*Peromyscus leucopus*) (Fahrig and Merriam 1985). Both modeling and empirical data supported the hypothesis that rodent populations in entirely or partly isolated patches (woodlots) have lower growth rates and therefore more easily go extinct than do populations in patches that are well connected. The differences were at least partly due to the greater possibility of establishment by dispersing animals in connected patches.

The problem of persistence is therefore strongly related to the problem of connectivity (Baudry and Merriam 1988; Merriam 1984; Taylor et al. 1993): "Connectivity is a parameter of landscape function which measures the processes by which subpopulations of organisms are interconnected into a functional demographic unit" (Merriam 1984). Connectivity is thus a concrete alternative to the straight-line dispersal distances usually implicitly assumed in island biogeography and metapopulation theories. Connectivity has been estimated in models as the probability of survival of animals moving between patches and therefore is not necessarily determined solely by the rate of dispersal (Henein and Merriam 1990). Knaapen et al. (1992) attempted to quantify connectivity according to the different resistances to dispersal exhibited by various habitats.

This operative definition of connectivity has led not only to extensive examination of dispersal among rodents but also to study of long-term persistence and breeding in the matrix and the potential use of corridors (Fig. 2.3) (Henderson et al. 1985; Merriam 1991; Merriam and Lanoue 1990; Middleton and Merriam 1981). Connectivity includes behavioral adaptations, in contrast to *connectedness* (Baudry and Merriam 1988), which refers only to physical linkages. For example, the winter lake landscape has high connectivity for certain large mammals (but usually not smaller ones or hibernators) that are able to disperse over ice (Lomolino 1982). Distinct barriers, such as roads (Mader 1984; Merriam et al. 1989), reduce connectivity to very low levels. Often the physical shelter of hedges, fencerows, riparian woodlands, and similar structures (corridors) is important for connectivity, but certain species have adapted to live at low densities and with large ranges in parts of the interstitial matrix. Thus, *P. leucopus,* which is known as a forest species in eastern North America, can live at lower densities but with larger movement ranges in cornfields in southern Ontario (Wegner and Merriam 1990) and can thereby link remnant forest patches without corridors. Also, other species utilize much larger ranges in heterogeneous than in homogeneous environments (Kozakiewicz 1995; Kozakiewicz and Szacki, Chap. 5; Szacki et al. 1993). Thus, animals sometimes make behavioral adaptations to fragmented landscapes, including the use of new habitats.

The persistence of subpopulations in habitat patches increases with increasing connectivity if corridors are of similar quality (Fahrig and Merriam 1985). However, low-quality corridors can, in simulation models, lead to lowered total populations because of higher mortality during movements between patches (Henein and Merriam 1990). Multipatch populations that depend on forcing individuals to move within a hostile matrix should thus be in peril of low persistence rates.

The concept of spatially variable connectivity has aroused further interest in landscape effects on the persistence, dispersion, and composition of patchily distributed species. Thus, van Apeldoorn et al. (1992) were able to show that local bank vole (*Clethrionomys glareolus*) populations in small forest fragments in agricultural landscapes in The Netherlands were dependent on the area of the nearest woodlots. The number of females decreased with the distance to permanently inhabited forests. Bowen (1982) and Kozakiewicz and Konopka (1991) also discovered genetic differences in patchily distributed low-density populations of *M. californicus* and between isolated and continuous populations of *C. glareolus,* respectively. These differences were considered to be due to founder effects. Merriam et al. (1989), however, were not able to demonstrate any genetic variation in subdivided populations of *P. leucopus.*

Work by Ilkka Hanski on shrew (*Sorex* spp.) persistence on lake islands has been based on a more direct theoretical framework in both island bio-

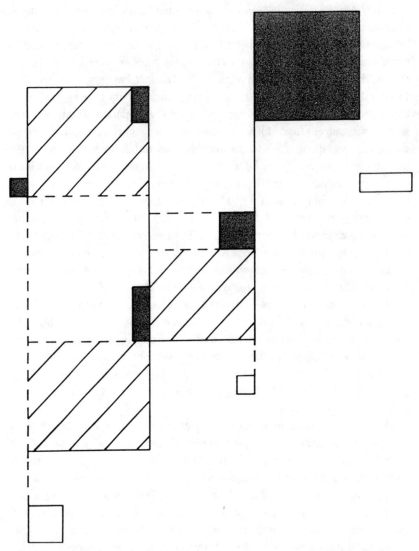

Fig. 2.3. A typical landscape, as considered with regard to the importance of connectivity for persistence. Habitat fragments (here small or large squares) may or may not be occupied, partly depending on the quality of corridors (uninterrupted or interrupted linear elements), but often also depending on the quality of the matrix (hatched or blank open space) as movement or breeding habitat.

geographic theory (Hanski 1986) and basic metapopulation theory (Peltonen and Hanski 1991). Both theories have been supported in these studies, which stressed the effects of distances to the mainland that supplied colonizing propagules of various species and of balanced extinction and colonization rates in separate species populations, partly derived from "incidence functions" (frequencies of occupation of various islands). Hanski and Kuitunen (1986) also demonstrated that the age structure of dispersers to such islands varied with density and with social, and probably also interspecific, interactions in the mainland populations. Thus, distant dynamics may affect the persistence of local small populations. In most studies of rodents, the importance of patch quality, corridors, and matrix composition has instead been stressed, and the random extinction and recolonization assumed in the classic metapopulation theory has thus not been well supported. The concept of metapopulation contains for some a strong connotation of randomness, and at least for rodents in habitat islands, it appears to be better to refer to patch (local) and regional populations.

Conservation Biology

The problem of the persistence of small populations is a pervasive issue in conservation biology. The first attempts to understand the implications of connectivity between woodlots for small mammals and birds (Henderson et al. 1985; Middleton and Merriam 1981; Wegner and Merriam 1979) were also directed toward general conservation problems.

The island biogeographic theory has been applied to conservation problems of large mammals in Africa (Soulé et al. 1979; Western and Ssemakula 1981) and to mammals in general in Canada (Glenn and Nudds 1989). The general conclusion has been that faunal collapses will occur in small reserves as a result of a lower equilibrium of species number due to higher extinction rates. However, Canadian generalist species seemed to survive well in small reserves, whereas most species adapted to undisturbed habitats disappeared. Not all the reserves were designed specifically for preserving the fauna; nevertheless, the general conflict between applications based on island biogeography and modern landscape thinking is obvious here. Glenn and Nudds (1989, Fig. 1b) suggested that the extinction rate of interior species increased in small reserves isolated in human-modified environments because large parts of these reserves, especially around the perimeter, were disturbed, and the remaining habitat could support only a few individuals. Further, they supposed that the immigration rate for this type of species was the same in large and small reserves. However, as the small reserves generally were surrounded by agricultural land or other disturbed environments, the connectivity for the habitat-specific species should have declined fairly abruptly, and immigration of new individuals should have been more or less impossible. Thus, the great difference between island biogeography and landscape ecology, at least as regards conservation, is that the former looks for equilibria in species numbers within reserves, while the latter mainly considers the importance of the surroundings (matrix and corridors) of the reserves for explaining movements of various species (see also Merriam 1993).

The use of corridors to bridge the gap between nature reserves is a conservation approach related to the concept of connectivity. Pros and cons have been discussed widely, for example, in relation to the Florida panther (*Felis concolor coryi*) (Harris and Gallagher 1989; Noss 1987; Simberloff and Cox 1987), but in general it still seems that very little is known about the general use of corridors by animals (Hobbs 1992; Saunders and Hobbs 1991; Simberloff et al. 1992). However, the corridors in these discussions have been actual physical connections, in the sense of connectedness as used by Baudry and Merriam (1988), whereas connectivity has wider implications, including consideration of various aspects of animal behavior in conservation biology (see following text for examples).

Rodents have been used to test the applicability of more or less formal metapopulation models in conservation (Ås et al. 1992). Examinations of the living conditions of certain regionally rare European mammals, such as squirrels (*Sciurus*) (Verboom and van Apeldoorn 1990) and badgers (*Meles*) (Lankester et al. 1991), have demonstrated the practical utility of connectivity measures in understanding persistence and extinction of local populations or clans. Similar observations have been made on assemblages of Australian terrestrial and arboreal marsupials (Bennett 1990; Laurance 1990, Chap. 3). However, the latter studies demonstrated great differences in connectivity (especially utilization of secondary vegetation) among the various marsupial species (Laurance 1990, 1991, Chap. 3). Habitat generalists demonstrated far greater connectivity than habitat specialists, and only specialists may need vegetation-specific corridors.

Perspectives

Landscape ecology has had a profound conceptual influence on mammalian population ecology during the last half century. Whereas population processes earlier were modeled on large, homogeneous areas, the emphasis has gradually shifted to the effects of spatial structure (Taitt and Krebs 1985, 593). Generally speaking, interest in all-pervading explanations for population regulation or limitation has waned, and instead local and multifactorial processes are being emphasized (Gaines et al. 1991; Lidicker 1988, 1991). The change has been less pronounced in mammalian community ecology, although attempts have been made to introduce landscape thinking at that level (Hansson and Henttonen 1988; Oksanen 1990).

Predator-prey relationships in heterogeneous landscapes appear to change with the relative proportion of the most productive habitat. Examples include both rodents and weasels (Oksanen 1990; Oksanen and Schneider, Chap. 7) and intermediate-sized mammals such as foxes and hares (Angelstam et al. 1984). An interest in gradients in predator-prey dynamics related to habitat juxtaposition (Akçakaya 1992; Pimm 1991) may thus signal changes even for community processes.

The importance of landscape composition for practical pest control, such as the eradication of refuge populations in order to avoid epizootic diseases and crop losses, was recognized long ago. The more recent theoretical advances in landscape ecology have led to the development of more elaborate pest control programs, but their applied use is obscure. In any case, they may have been more influential in preventive measures than in poisoning campaigns (Hansson 1992a; Redhead and Singleton 1988).

Development of conservation programs has relied strongly on empirical tests of ideas in island biogeography, metapopulation dynamics, and landscape ecology, and small mammals have often been used as test organisms. Certain practical applications for medium-sized mammals in forest fragments seem to have been implemented in densely populated and stressed environments in central Europe (Bissonette and Broekhuizen, Chap. 6).

Conservation work has demonstrated the need for thinking on a landscape scale (Fiedler and Jain 1992; Hansson 1992b). However, detailed studies, and especially experiments, are difficult to perform on large mammals in a landscape context. It has been suggested that model organisms be used to predict general landscape effects, and small mammals have been proposed as very suitable model organisms (Barrett et al., Chap. 8; Diffendorfer et al., Chap. 9; Ims and Stenseth 1989; N. C. Stenseth and R. A. Ims, pers. comm.). Such model studies will also provide more basic information on the behavior and dynamics of mammals in heterogeneous landscapes.

Summary

The consideration of landscapes instead of single habitat patches as basic environmental units for understanding mammalian ecology appears to have emerged first in pest control and was probably first articulated in the Soviet scientific literature. The concept has been incorporated into recent theoretical frameworks in Canada, Scandinavia, and the United States. Applications in pest control and conservation seem to have advanced particularly in Australia and some European countries. The general landscape ecology of mammalian populations has been elaborated according to two traditions, one stressing its importance for population dynamics and the other for population persistence. The former finds expression in applied work within pest control and the latter in similar work in conservation.

Work in population dynamics stresses the importance of adaptive and nonadaptive dispersal, the areal ratios and the relative quality of optimal and marginal patches, and the efficiency of different limiting mechanisms in various patches. In the latter connection, some authors stress the importance of social behavior and territoriality, and others the spatial interplay between predation and food limitation, especially species-specific nutritive adaptations and endurance of starvation.

The persistence of populations and communities has been examined from the point of view of island biogeographic theory, metapopulation dynamics, and landscape ecology, especially from a connectivity perspective. These approaches now seem to merge into a wider concept of landscape ecology. Connectivity remains a central factor but is broader than the idea of dispersal in population dynamics; it now also includes temporary persistence or breeding in both corridors and matrix. Community ecology of mammals has been reanalyzed in landscape contexts, and the importance of generalist predators in productive habitats for prey populations in adjoining areas has been stressed.

Recent concepts of landscape ecology have affected theory and research emphasis in basic mammalian ecology, while practical applications have been outlined but rarely implemented. The main impact appears to be increased awareness of local and temporary influences in population ecology as a replacement for single-factor explanations.

Acknowledgments

I am very grateful to William Z. Lidicker Jr., Gray Merriam, and John Wegner for information and comments.

Literature Cited

Anderson, P. K. 1970. Ecological structure and gene flow in small mammals. *Symp. Zool. Soc. London* 26:299–325.
Anderson, P. K. 1980. Evolutionary implications of microtine behavioral systems on the ecological stage. *The Biologist* 62:70–88.

Anderson, P. K. 1989. *Dispersal in Rodents: A Resident Fitness Hypothesis.* Spec. Publ. 9. American Society of Mammalogists. 141 pp.

Akçakaya, H. R. 1992. Population cycles of mammals. *Ecol. Monogr.* 62:120–42.

Angelstam, P., E. Lindström, and P. Widén. 1984. Role of predation in short-term population fluctuations of some birds and mammals in Fennoscandia. *Oecologia* 62:199–208.

Ås, S., J. Bengtsson, and T. Ebenhard. 1992. Archipelagoes and theories of insularity. Pages 201–51 *in* Hansson, L. (ed.), *Ecological Principles of Nature Conservation; Applications in Temperate and Boreal Environments.* Barking, England: Elsevier Applied Science. 436 pp.

Baudry, J., and H. G. Merriam. 1988. Connectivity and connectedness: Functional vs. structural patterns in landscapes. *Munstersche Geogr. Arb.* 29:23–8.

Bennett, A. F. 1990. Habitat corridors and the conservation of small mammals in a fragmented forest environment. *Landscape Ecol.* 4:109–22.

Bowen, B. S. 1982. Temporal dynamics of microgeographic structure of genetic variation in *Microtus californicus.* *J. Mammal.* 63:625–38.

Bujalska, G. 1990. Social system of the bank vole, *Clethrionomys glareolus.* Pages 155–68 *in* Tamarin, R. H., R. S. Ostfeld, S. R. Pugh, and G. Bujalska (eds.), *Social Systems and Population Cycles in Voles.* Basel, Switzerland: Birkhäuser Verlag. 229 pp.

den Boer, P. J. 1968. Spreading of risk and stabilization of animal numbers. *Acta Biotheor.* 18:165–94.

den Boer, P. J. 1981. On the survival of populations in a heterogeneous and variable environment. *Oecologia* 50:39–53.

Edler, A., and M. Mrciak. 1975. Gamasina mites (Acari:Parasitiformes) on small mammals in northernmost Fennoscandia. *Entomol. Tidskr.* 96:167–77.

Evans, F. C. 1942. Studies of a small mammal population in Bagley Wood, Berkshire. *J. Anim. Ecol.* 11:182–97.

Fahrig, L., and G. Merriam. 1985. Habitat patch connectivity and population survival. *Ecology* 66:1762–8.

Fiedler, P. L., and S. K. Jain (eds.). 1992. *Conservation Biology: The Theory and Practice of Nature Conservation, Preservation, and Management.* London: Chapman & Hall. 507 pp.

Foster, J., and M. S. Gaines. 1991. The effect of successional habitat mosaic on a small mammal community. *Ecology* 72:1358–73.

Frank, F. 1956. Grundlagen, Möglichkeiten und Methoden der Sanierung von Feldmausplagegebieten. *Nachrichtenbl. Dtsch. Pflanzenschutzdienstes (Braunschweig)* 8:147–58.

Gaines, M. S., N. C. Stenseth, M. L. Johnson, R. A. Ims, and S. Bondrup-Nielsen. 1991. A response to solving the enigma of population cycles with a multifactorial perspective. *J. Mammal.* 72:627–31.

Gilpin, M., and I. Hanski (eds.). 1991. *Metapopulation Dynamics: Empirical and Theoretical Investigations.* London: Academic Press. 336 pp.

Glenn, S. M., and T. D. Nudds. 1989. Insular biogeography of mammals in Canadian parks. *J. Biogeogr.* 16:261–8.

Hanski, I. 1986. Population dynamics of shrews on islands accord with the equilibrium model. *Biol. J. Linn. Soc.* 28:23–36.

Hanski, I., and J. Kuitunen. 1986. Shrews on small islands: Epigenetic variation elucidates population stability. *Holarctic Ecol.* 9:193–204.

Hansson, L. 1977. Spatial dynamics of field voles *Microtus agrestis* in heterogeneous landscapes. *Oikos* 29:539–44.

Hansson, L. 1990. Spatial dynamics in fluctuating vole populations. *Oecologia* 85:213–7.

Hansson, L. 1992a. Small mammal dispersal in pest management and conservation. Pages

177–98 *in* Stenseth, N. C., and W. Z. Lidicker Jr. (eds.), *Animal Dispersal; Small Mammals as a Model.* London: Chapman & Hall. 365 pp.

Hansson, L. (ed.). 1992b. *Ecological Principles of Nature Conservation; Applications in Temperate and Boreal Environments.* Barking, England: Elsevier Applied Science. 436 pp.

Hansson, L. 1994. Spatial dynamics in relation to density variations of rodents in a forest landscape. *Pol. Ecol. Stud.* 20:193-201.

Hansson, L., and H. Henttonen. 1988. Rodent dynamics as community processes. *Trends Ecol. Evol.* 3:195–200.

Harris, L. D., and P. B. Gallagher. 1989. New initiatives for wildlife conservation: The need for movement corridors. Pages 11–34 *in* Mackintosh, G. (ed.), *In Defense of Wildlife: Preserving Communities and Corridors.* Washington, D.C.: Defenders of Wildlife. 96 pp.

Henderson, M. T., G. Merriam, and J. Wegner. 1985. Patchy environments and species survival: Chipmunks in an agricultural mosaic. *Biol. Conserv.* 31:95–105.

Henein, K., and G. Merriam. 1990. The elements of connectivity where corridor quality is variable. *Landscape Ecol.* 4:157–70.

Hobbs, R. J. 1992. The role of corridors in conservation: Solution or bandwagon? *Trends Ecol. Evol.* 7:389–92.

Ims, R. A., and N. C. Stenseth. 1989. Divided the fruitflies fall. *Nature (London)* 342:21–2.

Knaapen, J. P., M. Scheffer, and B. Harms. 1992. Estimating habitat isolation in landscape planning. *Landscape Urban Plann.* 23:1–16.

Kozakiewicz, M. 1995. Resource tracking in space and time. Pages 136–48 *in* Hansson, L., L. Fahrig, and G. Merriam (eds.), *Mosaic Landscapes and Ecological Processes.* London: Chapman & Hall. 356 pp.

Kozakiewicz, M., and J. Konopka. 1991. Effect of habitat isolation on genetic divergence of bank vole populations. *Acta Theriol.* 36:363–8.

Lankester, K., R. van Apeldoorn, E. Meelis, and J. Verboom. 1991. Management perspectives for populations of the European badger (*Meles meles*) in a fragmented landscape. *J. Appl. Ecol.* 28:561–73.

Larsson, T.-B. 1975. Damage caused by small rodents in Sweden. *Ecol. Bull.* 19:47–56.

Laurance, W. F. 1990. Comparative responses of five arboreal marsupials to tropical forest fragmentation. *J. Mammal.* 71:641–53.

Laurance, W. F. 1991. Ecological correlates of extinction proneness in Australian tropical rain forest mammals. *Conserv. Biol.* 5:79–89.

Lefkovitch, L. P., and L. Fahrig. 1985. Spatial characteristics of habitat patches and population survival. *Ecol. Modell.* 30:297–308.

Levins, R. 1969. Some demographic and genetic consequences of environmental heterogeneity for biological control. *Bull. Entomol. Soc. Am.* 15:237–40.

Lidicker, W. Z., Jr. 1975. The role of dispersal in the demography of small mammals. Pages 103–28 *in* Golley, F. B., K. Petrusewicz, and L. Ryszkowski (eds.), *Small Mammals, Their Productivity and Population Dynamics.* Cambridge: Cambridge University Press. 451 pp.

Lidicker, W. Z., Jr. 1988. Solving the enigma of microtine "cycles." *J. Mammal.* 69:225–35.

Lidicker, W. Z., Jr. 1991. In defense of a multifactor perspective in population ecology. *J. Mammal.* 72:631–5.

Lomolino, M. V. 1982. Species-area and species-distance relationships of terrestrial mammals in the Thousand Island region. *Oecologia* 54:72–5.

Lomolino, M. V. 1986. Mammalian community structure on islands: Migration, extinction and interactive effects. *Biol. J. Linn. Soc.* 28:1–21.

MacArthur, R. H., and E. O. Wilson. 1967. *The Theory of Island Biogeography.* Princeton, N.J.: Princeton University Press. 203 pp.

Mader, H. J. 1984. Animal habitat isolation by roads and agricultural fields. *Biol. Conserv.* 29:81–96.

Merriam, H. G. 1984. Connectivity: A fundamental ecological characteristic of landscape pattern. *Proc. Int. Semin. Methodol. Landscape Ecol. Res. Plann.* 1:5–15.

Merriam, G. 1991. Corridors and connectivity: Animal populations in heterogeneous environments. Pages 133–42 *in* Saunders, D., and R. Hobbs (eds.), *Nature Conservation 2: The Role of Corridors.* Chipping Norton, N.S.W., Australia: Surrey Beatty & Sons. 442 pp.

Merriam, G. 1993. Managing the land: A medium-term strategy for integrating landscape ecology into environmental research and management. Report to the Ontario Ministry of Natural Resources, Sault St. Marie. 39 pp.

Merriam, G., M. Kozakiewicz, E. Tsuchiya, and K. Hawley. 1989. Barriers as boundaries for metapopulations and demes. *Landscape Ecol.* 2:227–35.

Merriam, G., and A. Lanoue. 1990. Corridor use by small mammals: Field measurement for three experimental types of *Peromyscus leucopus. Landscape Ecol.* 4:123–31.

Merriam, G., and J. Wegner. 1992. Local extinctions, habitat fragmentation, and ecotones. Pages 150–69 *in* Hansen, A. J., and F. di Castri (eds.), *Landscape Boundaries; Consequences for Biotic Diversity and Ecological Flows.* Ecological Studies 92. New York: Springer-Verlag. 452 pp.

Middleton, J. D., and G. Merriam. 1981. Woodland mice in a farmland mosaic. *J. Appl. Ecol.* 18:703–10.

Middleton, J. D., and G. Merriam. 1983. Distribution of woodland species in farmland woods. *J. Appl. Ecol.* 20:625–44.

Naumov, N. P. 1936. (On some peculiarities of ecological distribution of mouse-like rodents in southern Ukraine.) *Zool. Zh.* 15:675–96.

Naumov, N. P. 1948. (Sketches of the Comparative Ecology of Mouse-like Rodents.) Moscow: Izd. Akad. Nauk.SSR. 202 pp.

Naumov, N. P. 1972. *The Ecology of Animals.* Urbana: University of Illinois Press. 650 pp.

Newsome, A. E. 1969a. A population study of house-mice temporarily inhabiting South Australian wheatfields. *J. Anim. Ecol.* 38:341–59.

Newsome, A. E. 1969b. A population study of house-mice permanently inhabiting a reed bed in South Australia. *J. Anim. Ecol.* 38:361–77.

Noss, R. F. 1987. Corridors in real landscapes: Reply to Simberloff and Noss. *Conserv. Biol.* 1:159–64.

Oksanen, T. 1990. Exploitation ecosystems in heterogeneous habitat complexes. *Evol. Ecol.* 4:220–34.

Ostfeld, R. S. 1992a. Small-mammal herbivores in a patchy environment: Individual strategies and population responses. Pages 43–74 *in* Hunter, M. D., T. Ohgushi, and P. W. Price (eds.), *Effects of Resource Distribution on Animal-Plant Interactions.* New York: Academic Press. 505 pp.

Ostfeld, R. S. 1992b. Effects of habitat patchiness on population dynamics: A modelling approach. Pages 851–63 *in* McCullough, D. R., and R. H. Barrett (eds.), *Wildlife 2001: Populations.* London: Elsevier Applied Science. 1,163 pp.

Ostfeld, R. S., and L. L. Klosterman. 1986. Demographic substructure in a California vole population inhabiting a patchy environment. *J. Mammal.* 67:693–704.

Ostfeld, R. S., W. Z. Lidicker Jr., and E. J. Heske. 1985. The relationship between habitat heterogeneity, space use, and demography in a population of California voles. *Oikos* 45:433–42.

Peltonen, A., and I. Hanski. 1991. Patterns of island occupancy explained by colonization and extinction rates in shrews. *Ecology* 72:1698–708.

Pimm, S. L. 1991. *The Balance of Nature? Ecological Issues in the Conservation of Species and Communities.* Chicago: University of Chicago Press. 434 pp.

Ray, C., M. Gilpin, and A. T. Smith. 1991. The effect of conspecific attraction on metapopulation dynamics. *Biol. J. Linn. Soc.* 42:123–34.

Redhead, T. D. 1988. Prevention of plagues of house mice in rural Australia. Pages 191–205 in Prakash, I. (ed.), *Rodent Pest Management.* Boca Raton, Fla.: CRC Press. 480 pp.

Redhead, T. D., N. Enright, and A. E. Newsome. 1985. Causes and predictions of outbreak of *Mus musculus* in irrigated and non-irrigated cereal farms. *Acta Zool. Fenn.* 173:123–7.

Redhead, T., and G. Singleton. 1988. The PICA strategy for the prevention of losses caused by plagues of *Mus domesticus* in rural Australia. *EPPO Bull.* 18:237–48.

Risser, P. G., J. R. Karr, and R. T. T. Forman. 1984. *Landscape ecology; Directions and approaches.* Spec. Publ. 2. Champaign: Illinois Natural History Survey. 18 pp.

Saunders, D. A., and R. J. Hobbs. 1991. *Nature Conservation 2: The Role of Corridors.* Chipping Norton, N.S.W., Australia: Surrey Beatty & Sons. 442 pp.

Simberloff, D., and J. Cox. 1987. Consequences and costs of conservation corridors. *Conserv. Biol.* 1:63–71.

Simberloff, D., J. A. Farr, J. Cox, and D. W. Mehlman. 1992. Movement corridors: Conservation bargains or poor investments? *Conserv. Biol.* 6:493–504.

Singleton, G. R. 1989. Population dynamics of an outbreak of house mice (*Mus domesticus*) in the mallee wheatlands in Australia—Hypothesis of plague formation. *J. Zool.* 219:495–515.

Smith, A. T. 1974. The distribution and dispersal of pikas: Consequences of insular population structure. *Ecology* 55:1112–9.

Soulé, M. E., B. A. Wilcox, and C. Holtby. 1979. Benign neglect: A model of faunal collapse in the game reserves of East Africa. *Biol. Conserv.* 15:259–72.

Stenseth, N. C. 1977. On the importance of spatio-temporal heterogeneity for the population dynamics of rodents: Towards a theoretical foundation of rodent control. *Oikos* 29:545–52.

Stenseth, N. C. 1980. Spatial heterogeneity and population stability: Some evolutionary consequences. *Oikos* 35:165–84.

Stenseth, N. C. 1981. How to control pest species: Applications of models from the theory of island biogeography in formulating pest control strategies. *J. Appl. Ecol.* 18:773–94.

Stenseth, N. C., and L. Hansson. 1981. The importance of population dynamics in heterogeneous landscapes: Management of vertebrate pests and some other animals. *Agro-Ecosystems* 7:187–211.

Stickel, L. F. 1979. Population ecology of house mice in unstable habitats. *J. Anim. Ecol.* 48:871–87.

Szacki, J., J. Babińska-Werka, and A. Liro. 1993. The influence of landscape spatial structure on small mammal movements. *Acta Theriol.* 38:113–24.

Taitt, M. J., and C. J. Krebs. 1985. Population dynamics and cycles. Pages 567–620 in Tamarin, R. H. (ed.), *Biology of New World Microtus.* Pittsburgh: American Society of Mammalogists. 893 pp.

Taylor, P. D., L. Fahrig, K. Henein, and G. Merriam. 1993. Connectivity is a vital element of landscape structure. *Oikos* 68:571–3.

van Apeldoorn, R. C., W. T. Oostenbrink, A. van Winden, and F. F. van der Zee. 1992. Effects of habitat fragmentation on the bank vole, *Clethrionomys glareolus,* in an agricultural landscape. *Oikos* 65:265–74.

Verboom, J., and R. van Apeldoorn. 1990. Effects of habitat fragmentation on the red squirrel, *Sciurus vulgaris* L. *Landscape Ecol.* 4:171–6.

Wegner, J. F., and G. Merriam. 1979. Movements by birds and mammals between a wood and adjoining farmland habitats. *J. Appl. Ecol.* 16:349–57.
Wegner, J. F., and G. Merriam. 1990. Use of spatial elements in a farmland mosaic by a woodland rodent. *Biol. Conserv.* 54:263–76.
Western, D., and J. Ssemakula. 1981. The future of the savannah ecosystem: Ecological islands or faunal enclaves? *Afr. J. Ecol.* 19:7–19.

Part II

Field Approaches: Evidence and
Contributions to Theory

Field research is at the heart of investigations in landscape ecology. We must know what real landscapes are like and observe how organisms interact with them. This is a necessary first step in asking process-oriented questions that may themselves be amenable to experimental manipulations. It is also critical to study natural landscapes and compare them to human-made mosaics. Such comparisons can provide insights into the kind of traits that lead to success or failure under increasing fragmentation. Organisms must also be investigated within their communities, or we will fail to discover how various species' interactions (coactions) affect their response to landscape elements, both traditional and novel.

The five chapters in Part II sample a diversity of field approaches, illustrating both their power and their limitations. Substantial contributions to both theory and the empirical database are demonstrated.

Bill Laurance (Chap. 3) summarizes some aspects of his long-term studies in the tropical rain forest of northern Queensland. Most of the forest fragments that he has studied are 75 or more years old, so the mammalian communities have had a chance to equilibrate. This is a very important advantage over other tropical investigations, such as in South America, that involve recently formed forest fragments. The non-volant fauna is composed of various rodents and marsupials ranging in size up to tree-kangaroos. Laurance's analysis focuses on edge effects and life history features that promote or deter extinction in forest fragments.

The chapters by Gray Merriam (Chap. 4) and Michał Kozakiewicz and Jakub Szacki (Chap. 5) discuss the movements of small mammals in strongly subdivided agricultural and urban landscapes. They are concerned with how different species use corridors and other matrix elements to maintain a metapopulation structure. Species lacking good dispersal capabilities are extinction-prone under these conditions. Both chapters raise the issue of to what extent good dispersers were already that way before anthropogenic fragmentation and to what extent they are merely responding to recent selection pressures. Gray explains the notion of *operational demographic unit* and argues that this concept is important if we are to understand population dynamics in fragmented landscapes. Michał and Jakub raise the possibility that the basic home range concept, upon which so much of our population theory is based, may be an inappropriate model for strongly fragmented landscapes.

Chapter 6 is by two authorities on the ecology of martens (*Martes*), John Bissonette and Sim Broekhuizen. In a data-rich contribution, they combine their extensive knowledge of three species—one North American and two European—and attempt to understand how these medium-sized mammals cope with habitat fragmentation. They also explore insights gained from viewing the problem on various spatial scales and conclude that a multiscale approach is essential.

The final chapter in this part also deals with carnivores, but in the context of predator-prey relationships. Tarja Oksanen and Michael Schneider first review the development of theory regarding trophic relations in communities with different numbers of trophic levels, the number of levels being determined by the primary productivity of the community. If predators are present, herbivore numbers tend to be regulated by the predators and not by their own food supplies. Then Tarja and Michael explore how these trophic relations are affected by heterogeneous environments in which the patches exhibit different primary productivities. Finally, they illustrate these new ideas with their own data from arctic subalpine habitats in northern Norway. Clearly, the ratio of productive to less productive habitat patches in the landscape is critical to the impacts of predation, and this insight connects us back to the discussion in Chapter 1 about microtine demography in landscapes with different ratios of optimal habitat.

A landscape in northern Norway (Joatka, Finnmark) relatively unmodified by human influences (see Chap. 7). (Photo by T. Oksanen, taken in September 1986.)

3

Extinction and Survival of Rainforest Mammals in a Fragmented Tropical Landscape

William F. Laurance

In Australia, tropical rainforests are confined to a series of disjunct tracts skirting the eastern margin of Cape York Peninsula (Fig. 3.1). These ancient, relict forests make up only a tiny fraction (< 0.5%) of the continent's land area, yet they sustain the greatest species richness and endemism of any terrestrial habitat in Australia (Keto and Scott 1986).

Australia is now recognized as a world leader in conservation of its tropical rainforest. In 1988, following years of rancorous debate between conservationists and logging interests, the largest rainforest tract (situated between Townsville and Cooktown [Fig. 3.1]) was formally designated the Wet Tropics of Queensland World Heritage Area. This 900,000 ha complex of rainforest, mangroves, and other tropical habitats is now rigorously protected, with a moratorium on nearly all logging, clearing, and mining activities (WTMA 1992).

In many ways, however, this enlightened stance belies the history of land use in this region. Early European settlers in Australia were quick to exploit the gold and rich timber resources of north Queensland and were soon followed by farmers and pastoralists (Hughes 1987). Today, most lowland rainforests south of the Daintree River have been cleared for sugarcane, while the rainforests of the Atherton Tableland have been nearly denuded for cattle pastures (Fig. 3.1). In addition to widespread clearing, most accessible areas of rainforest were subjected to intensive selective logging (Laurance 1986).

From the perspective of faunal conservation, the loss and fragmentation of upland forests has been particularly alarming. It is currently accepted that during repeated cooling and drying episodes in the Pleistocene, Australian tropical rainforests shrank into a series of small, isolated refugia, usually located in hyperwet upland areas (Hopkins et al. 1993; Kershaw 1986; Webb

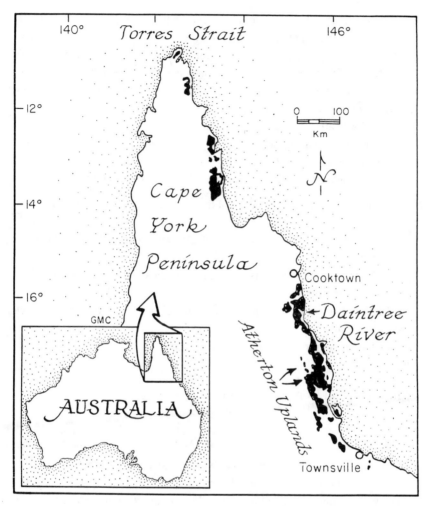

Fig. 3.1. The current distribution of tropical rainforest (shaded areas) in Australia. The Atherton Uplands region is a key center of endemism for nonflying mammals. (Reprinted from Laurance 1990a; courtesy of the American Society of Mammalogists.)

and Tracey 1981). Today, these upland sites are crucial centers of endemism and diversity for many taxa (Laurance 1987). The Atherton Uplands, for example, sustain the highest diversity of nonflying mammals of any region in Australia (Laurance 1993a; Winter et al. 1984).

Given its unquestioned biological significance and the fact that most of the area's forest fragments were formed by the early 1920s (Frawley 1983), the Atherton Tableland provides a vital natural laboratory for studies of rainforest fragmentation. Work to date has focused on assessing the effects of fragmentation on nonflying mammals (Laurance 1987, 1989, 1990a, 1991a, 1993a, 1994; Pahl et al. 1988) and birds (Crome et al. 1995; Laurance et al. 1993; Wharburton 1987); determining the importance of edge effects and wind disturbance in fragments (Laurance 1991b; Laurance and Yensen 1991); and studying genetic variation in fragmented vertebrate populations (Leung et al. 1993; C. Moritz and L. Joseph, unpub. data).

In this chapter I highlight key aspects of these investigations, emphasizing my own studies of the ecology of nonflying mammals and their habitats within this heavily fragmented landscape. These studies had five goals:

1. To determine the responses of native mammal species to forest fragmentation;

2. To assess the importance of edge effects and other ecological changes in fragments;

3. To identify ecological traits of mammal species that influence their response to fragmentation;

4. To identify fragment and landscape features that affect mammal species assemblages; and

5. To make recommendations for the design of faunal corridors and nature reserves.

Study Area and Census Methods

Study Area

The principal study area was the southern half of the Atherton Tableland (Fig. 3.2), a hilly, midelevation (600–900 m) plateau in northeastern Queensland. Most rainforests in this vicinity are classified as complex notophyll vine forest, with a complex structure, multiple tree layers, and many epiphytes, lianas, and climbing rattans (Tracey 1982).

Large-scale clearing of Tableland forests began about 1909 and proceeded rapidly for the next three decades (Frawley 1983). By 1983, more than 76,000 ha of forest had been removed (Winter et al. 1987), leaving more than 100 forest fragments, ranging from 1 to 600 ha in area, scattered over an area of about 900 km². Large (> 3,000 ha) forest tracts survive only on the steep hillsides that nearly encircle the Tableland. Most fragments are surrounded by cattle pastures interspersed with narrow (10–50 m wide), often discontinuous strips of forest regrowth along streams (Fig. 3.2). Pastures range from grass monocultures to mixed grasslands with weedy forbs and shrubs.

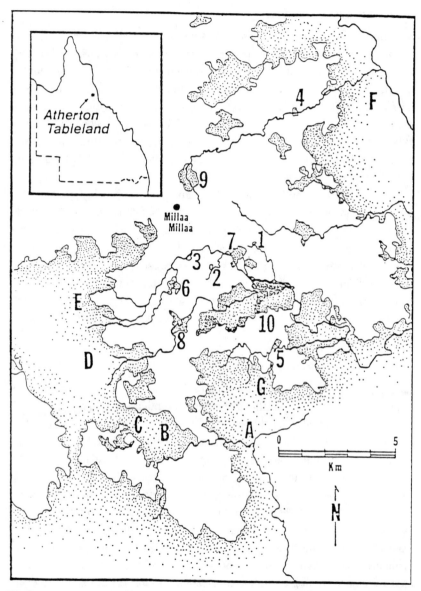

Fig. 3.2. Map of the principal study area on the southern Atherton Tableland. Capital letters denote control sites; the numbers (1–10) show the locations of forest fragments (1.4–590 ha in area). Unstippled areas are mostly cattle pastures. Dark, wavy lines are streams or rivers. (Reprinted from "Ecological correlates of extinction proneness in Australian tropical rain forest mammals" by W. F. Laurance (1991a), *Conservation Biology*, vol. 5, pp. 79–89; reprinted by permission of the Society for Conservation Biology and Blackwell Scientific Publications, Inc.)

Census Methods

Mammal populations were intensively surveyed in 1984 and 1986–87; additional ecological studies were conducted from 1989 to 1994. Arboreal species were censused with spotlights at night in seven control sites in unfragmented forest, 10 forest fragments ranging from 1.4 to 590 ha in area, and three corridors of regrowth along streams (Fig. 3.2). Each site was censused seven to nine times, for a total of 1,062 observations of five arboreal species (Laurance 1990a). Data on numbers of the red-legged pademelon (*Thylogale stigmatica*), a small rainforest wallaby, also were collected during the spotlighting (230 detections).

Four trapping methods were used to census small (< 4 kg) rodents and marsupials: ground-set, arboreal, predator, and pitfall traps. Standard mark-recapture methods were used except for pitfall captures. Traps were placed in small (900 m²) grids, with 32 grids in the 10 fragments and 20 grids in five of the control sites. Small mammals also were censused in three pastures and three corridors with 12 transects. Each grid or transect was trapped for five consecutive nights on six or seven occasions. In 22,580 trap-nights of effort, 6,382 captures of 17 small mammal species were recorded (Laurance 1990b, 1994).

Ecological Changes in Rainforest Fragments

Rainforest fragments on the Atherton Tableland are not unaltered samples of pristine rainforest, but rather have been modified by an array of edge effects and substantial wind disturbance (Fig. 3.3). Relative to control sites, forest fragments smaller than 600 ha typically have reduced canopy cover, more fallen trees and snapped boles, and exceptional abundances of disturbance-adapted plants (e.g., lianas, climbing rattans [*Calamus* spp.], stinging trees [*Dendrocnide* spp.], and the grass *Oplismenus hirtellus*) and exotic plants (e.g., *Rubus alcefolius, Solanum dallachii, S. hamulosum,* and *S. mauritianum*) (Laurance 1991b, unpub. data). These disturbances are evident at least 500 m into the forest from fragment margins (Laurance 1991b) and probably occur when powerful wind shear forces, which are readily generated over surrounding denuded landscapes (Hobbs 1993a; Saunders et al. 1991), strike abrupt fragment edges.

In addition, forest fragments on the Atherton Tableland (Fig. 3.4) are subjected to a "seed rain" of weedy propagules from exotic and secondary plants growing in surrounding modified habitats (Willson and Crome 1989). By colonizing treefall gaps, these species may progressively alter the floristic com-

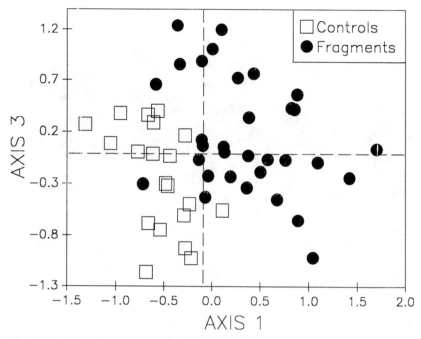

Fig. 3.3. Ranking of forest-structural and floristic attributes of 32 study plots in forest fragments and 20 control plots by nonmetric multidimensional scaling. Plots with higher values on axis 1 had more treefalls, reduced canopy cover, and elevated abundances of gap-
colonizing (*Dendrocnide* spp., *Calamus* spp.) and exotic (*Solanum* spp.) plants. Plots with
higher values on axis 3 had reduced subcanopy cover and more lianas. Fragment plots
were significantly skewed toward higher values on both axes, illustrating the tendency of
fragments to be "hyperdisturbed." (Reprinted from *Biological Conservation*, vol. 69, W. F.
Laurance, "Rainforest fragmentation and the structure of small mammal communities in
tropical Queensland," pp. 23–32, copyright 1994, with kind permission from Elsevier Science Ltd., The Boulevard, Langford Lane, Kidlington 0X5 1GB, U.K.)

position of fragments (Janzen 1983). Finally, forest edges and disturbed forest exhibit a series of microclimatic changes, such as greater variation in air
temperature, higher insolation, and reduced soil moisture and relative hu-
midity (Kapos 1989), that clearly affect some specialized plant and animal
species (Lovejoy et al. 1986).

Rainforest Mammals

Forest Rodents

Four species of rainforest rodents—the fawn-footed melomys (*Melomys cervinipes*), Cape York rat (*Rattus leucopus*), bush rat (*R. fuscipes*), and white-

Fig. 3.4. A typical landscape on the Atherton Tableland (4 km south of Millaa Millaa), showing a small (6.3 ha) wind-disturbed rainforest fragment surrounded by a "sea" of cattle pastures. (Photo by W. F. Laurance, taken in 1987.)

tailed rat (*Uromys caudimaculatus*)—are very abundant in north Queensland and accounted for the large majority (> 98%) of small-mammal captures (Laurance 1994). These rodents apparently have a strong influence on forest dynamics as major predators of seeds and fruits (Osunkoya 1994) and of insects and small vertebrates (Laurance et al. 1993; Laurance and Grant 1994) and as important prey for owls, pythons (*Morelia* spp.), and spotted-tailed quolls (*Dasyurus maculatus*).

The four rodents exhibited complex distribution patterns within the study area (Laurance 1994). I summarize here the major trends.

1. All four species were abundant in at least some fragments, and all were regularly captured within the matrix of pastures and regrowth forest surrounding fragments.

2. Three of the four rodents (the melomys, Cape York rat, and bush rat) clearly were influenced by forest edges or disturbed forest near edges (i.e., they exhibited significant positive or negative correlations with variables quantifying forest disturbance or the proximity of study sites to forest edges). Thus, ecological changes in fragments played an important role in structuring rodent assemblages.

3. Overall, rodents were significantly more abundant in both small (1.4–12.7 ha) and large (21–590 ha) fragments than in control sites. This may have resulted from habitat changes in fragments (such as edge conditions and increased levels of disturbance) that were generally favorable to rodents, from novel foraging opportunities in surrounding pastures, or from a decline of rainforest-dependent predators or competitors in fragments.

4. Experimental studies suggest that elevated rodent populations in fragments could increase rates of predation on rainforest seeds and fruits (Osunkoya 1994) and on eggs of ground-nesting birds (Laurance et al. 1993).

5. Competitive interactions were important in structuring rodent assemblages, especially for the Cape York rat and bush rat, which were strongly and negatively correlated in abundance. These interactions were strongest in fragments, where the resource base was limited, which suggests that interspecific competition may help drive extinctions of ecologically similar species in fragments (Gilpin and Diamond 1982; Grant 1966).

Trappable Rare Species

Six species of rare, forest-dependent marsupials or rodents were captured infrequently (one to 16 times each), despite the diversity of trapping methods employed (Laurance 1992): the mainly insectivorous brown antechinus (*Antechinus stuartii*), yellow-footed antechinus (*A. flavipes*) (Fig. 3.5a), Atherton antechinus (*A. godmani*), and white-footed dunnart (*Sminthopsis leucopus*); the primitive musky rat-kangaroo (*Hypsiprimnodon moschatus*); and the carnivorous water rat (*Hydromys chrysogaster*). The following trends were detected (Laurance 1990b, 1994).

1. The responses of these rare species to fragmentation varied. Three declined significantly in fragments (the brown and Atherton antechinuses and the musky rat-kangaroo), one did not differ significantly between controls and fragments (the water rat), and one increased significantly in fragments (the yellow-footed antechinus). The sample size for the white-footed dunnart was too small to permit statistical analysis, but it is likely to respond negatively to fragmentation (Laurance 1990b), given its strong dependence on rainforest in north Queensland (Van Dyck 1985).

2. Despite these individualistic responses, on an overall basis, rare forest mammals declined significantly in abundance in fragments, relative to control sites.

3. Among these six species, only the water rat and yellow-footed antechi-

Fig. 3.5. Three Australian rainforest marsupials. (a) The yellow-footed antechinus *(Antechinus flavipes)*, a small marsupial predator that favors fragment margins (photo by M. P. Trenerry). (b) The green ringtail possum *(Pseudocheirops archeri)*, a folivore endemic to rainforest that becomes superabundant in regrowth forest (photo by W. F. Laurance). (c) The Herbert River ringtail possum *(Pseudocheirus herbertensis)*, a rainforest endemic with a restricted upland distribution; this animal declines in smaller (< 20 ha) fragments and rarely uses stream corridors (photo by M. P. Trenerry).

nus were captured in the matrix. In rainforest, the yellow-footed antechinus was usually captured within 35 m of forest edges.

4. The spotted-tailed quoll (*Dasyurus maculatus*), a medium-sized (4–10 kg) marsupial predator, was never encountered despite intensive efforts to capture it. Local residents commonly observed this distinctive mammal before World War II (Laurance 1989), but it now appears to be extremely rare on the Atherton Tableland. The quoll occurs at low population densities and may be severely affected by forest loss and fragmentation, compounded by widespread invasions of cane toads (*Bufo marinus*), which produce potent skin toxins and are lethal when consumed by native predators (Covacevich and Archer 1970).

Grassland and Introduced Species

Six species of mammals were usually captured in pastures, along forest-pasture edges, or near human habitations: the long-nosed bandicoot (*Perameles nasuta*) and northern brown bandicoot (*Isoodon macrourus*); two grassland rodents, the swamp rat (*Rattus lutreolus*) and cane rat (*R. sordidus*); and the introduced black rat (*R. rattus*) and house mouse (*Mus musculus*). The following trends were noted (Laurance 1994).

1. House mice and swamp rats were trapped frequently in pastures, but each was captured only once in fragments. These and other nonforest rodents probably are excluded from fragments and control sites by abundant and aggressive rainforest rodents.

2. Black rats were captured only within 200 m of human habitations. None were encountered in rainforest.

3. The long-nosed bandicoot was captured occasionally in deep rainforest but was most often captured along stream corridors, near forest edges, and in adjoining pastures.

Arboreal Mammals

Five species of arboreal, leaf-eating marsupials were censused by spotlighting: the coppery brushtail possum (*Trichosurus vulpecula*); three species of tropical ringtail possum, the green ringtail (*Pseudocheirops archeri*) (Fig. 3.5b), Herbert River ringtail (*Pseudocheirus herbertensis*) (Fig. 3.5c), and lemuroid ringtail (*Hemibelideus lemuroides*); and Lumholtz's tree-kangaroo (*Dendrolagus lumholtzi*). Censuses in fragments, controls, and stream corridors revealed the following trends (Laurance 1990a).

1. Arboreal species exhibited a strong gradient in extinction proneness. Most vulnerable were lemuroid ringtails, which declined by over 97% in fragments and corridors. Herbert River ringtails and tree-kangaroos exhibited negative but intermediate responses, whereas coppery brushtails and green ringtails were least affected, relative to populations in controls.

2. For the most vulnerable species, the pace of extinction was rapid. The lemuroid ringtail, for example, is known to have disappeared from a small (1.4 ha) fragment in only three to nine years and from larger (43–75 ha) fragments in 35–60 years (Laurance 1990a).

3. Two of the most vulnerable species, the lemuroid and Herbert River ringtails, avoided or very rarely used corridors of regrowth along streams. The green ringtail, however, actually became superabundant in corridors.

Ecological Correlates of Extinction Proneness

Why did some species disappear or decline in fragments, whereas others remained stable or even increased? To address this question, I tested the utility of seven ecological traits for predicting the responses to fragmentation of 16 forest-dependent species. The seven traits were body size, longevity, fecundity, trophic level, degree of dietary specialization, abundance in unfragmented rainforest, and matrix tolerance (i.e., the relative abundance of each species in the modified habitats surrounding fragments).

Surprisingly, matrix tolerance emerged as an overriding correlate of vulnerability (Fig. 3.6). Species that tolerated or exploited the matrix often remained stable or increased in fragments, whereas those that avoided these habitats declined or disappeared (Laurance 1990a, 1991a, 1994). There are three probable reasons why matrix tolerance was such an effective predictor of species vulnerability.

1. Species that use the matrix are much better at dispersing between fragments or between "mainland" areas and fragments. The demographic and genetic contributions of immigrants can bolster small, dwindling populations in fragments and provide a buffer against extinction (the "rescue effect" [Brown and Kodric-Brown 1977]). Matrix-tolerant species also can recolonize fragments following the extinction of local populations (Merriam, Chap. 4).

2. Matrix-tolerant species are commonly preadapted for ecological changes in fragments, such as edge effects. For example, the coppery brushtail and green ringtail possums, both common in the matrix, feed on secondary trees (especially *Alphitonia petrei*) that proliferate along corri-

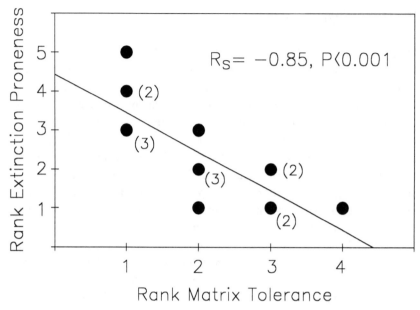

Fig. 3.6. Effect of matrix tolerance on extinction proneness of 16 forest-dependent mammal species in tropical Queensland (Spearman rank correlation). (Reprinted from "Ecological correlates of extinction proneness in Australian tropical rain forest mammals" by W. F. Laurance (1991a), *Conservation Biology,* vol. 5, pp. 79–89; reprinted by permission of the Society for Conservation Biology and Blackwell Scientific Publications, Inc.)

dors and fragment margins. In addition to being effective dispersers, these folivores respond positively to edge conditions in fragments.

3. Species that have declined regionally (such as the spotted-tailed quoll) would tend to be absent from both fragments and the matrix. This would strengthen the correlation between matrix tolerance and extinction proneness.

Interestingly, the abundance of species in control sites was often a misleading predictor of vulnerability. For example, the lemuroid ringtail was the most abundant arboreal mammal in controls yet clearly was the most extinction-prone. Likewise, the green ringtail, overall the rarest arboreal species in controls, was common in fragments and superabundant in corridors. Although natural rarity has often been invoked as a factor influencing survival of vertebrates in insular communities (Diamond 1984; Terborgh 1975; Wilcox 1980), it may be an effective indicator only for those species that avoid the matrix (Diamond et al. 1987). In this regard, islands are probably poor analogues of habitat fragments, because the matrix is inhospitable (Laurance 1991a).

Reserve Design Considerations

Landscape Predictors of Arboreal Mammal Richness

Because they depend strongly on forest vegetation for movement, arboreal mammals may be influenced by the nature of matrix habitats surrounding forest fragments. On the Atherton Tableland, fragments vary considerably in their degrees of spatial isolation and "connectedness" (i.e., the extent to which each is linked by well-vegetated stream corridors to mainland forest tracts [Merriam, Chap. 4]). I devised six statistically independent predictors that quantified key landscape features of the fragments in my study area—area, shape, age, topographic diversity, spatial isolation, and index of corridor quality—and used stepwise multiple regression analysis to assess the effects of these variables on arboreal species richness.

The model selected three variables (fragment area, corridor quality, and spatial isolation) that yielded a highly significant multiple regression (F = 12.85, R^2 = 82%, df = 8, P = 0.005). Fragment area explained 42% and the other two variables accounted for 40% of the variation in species richness. Thus, for arboreal mammals, the degree of connectedness and spatial isolation of fragments appeared nearly as important as fragment area in predicting species richness (Laurance 1990a).

Design of Faunal Corridors

Despite concerns about potential risks of faunal corridor systems (Simberloff and Cox 1987; Simberloff et al. 1992), many theoretical (Burkey 1989; Fahrig and Merriam 1985; Henein and Merriam 1990) and empirical (Bennett 1990a,b; Fahrig and Merriam 1985; Harris 1984; Kozakiewicz and Szacki, Chap. 5; Merriam, Chap. 4; Saunders and de Rebeira 1991) studies suggest that movement of individuals between discrete habitat patches promotes survival of wildlife populations. In this study, strong circumstantial evidence suggested that linkages provided by regrowth along streams enhanced mammalian richness in fragments, apparently by facilitating dispersal of animals among and between fragments and large forest tracts (Laurance 1990a, 1991a). These riparian strips also provide habitat and food resources for some species (Crome et al. 1995). The following points summarize my views on faunal corridors in this region (see also Laurance 1991a).

1. Because of their elongated shape, corridors are highly prone to edge effects, wind disturbance, and invasions of exotic and domestic species (e.g.,

weeds, cane toads, feral pigs, dogs, cats, and cattle). All but the widest
(> 1 km) corridors are likely to differ from primary forest in some aspects
of forest structure, composition, and microclimate.

2. Corridors should be explicitly designed to accommodate the needs of
rainforest specialists, which are often highly sensitive to fragmentation, in
large part because they avoid the modified habitats that constitute the matrix. Many specialists also avoid narrow (< 100 m) corridors and those composed only of regrowth (Bissonette and Broekhuizen, Chap. 6). Poorly designed corridors will act as "selective filters," permitting movement of more
generalized species but not many rainforest specialists.

3. Corridors are often created by reforestation. Reforestation of rainforest is a slow process but may be accelerated by planting of mature-phase
rainforest trees under a subcanopy of robust pioneer species (Tracey 1986)
and by the use of fruiting pioneer species (such as *Omalanthus novo-
guineenensis* [Laurance 1993b]) that are highly attractive to rainforest
frugivores and seed dispersers.

4. Stream gullies are good locations for corridors, for four reasons: gullies are less exposed to wind disturbance; many wildlife species require regular access to water; vegetated streams facilitate movement of aquatic or
semiaquatic animals; and gullies retain moisture and often contain deep,
alluvial soils, which promote plant growth and thus speed successional
processes.

5. Corridors may require fences or active protection from feral or domestic animals. Edge-inhabiting predators may be problematic in some corridors (Ambuel and Temple 1983; Simberloff and Cox 1987).

6. Long (> 5 km) corridors may be prohibitively expensive to create and
maintain (Hobbs 1993b). Fragment length and width probably interact, with
longer corridors requiring greater minimum widths than shorter corridors.

Design of Tropical Reserve Systems

I close this essay with four points about the design of nature reserves in tropical communities.

1. Tropical rainforests typically have high beta and gamma diversity
(Erwin 1986; Foster 1980). Consequently, a single megareserve will fail to
capture much of the biological diversity of a region. A better strategy is to
stratify a number of small to medium-sized reserves across major environmental gradients (e.g., elevation and moisture). These smaller reserves are
designed to maximize the conservation of habitat diversity and to preserve

populations of localized endemics, patchily distributed species, and rare vegetation types.

2. Such a regional strategy should include at least one large reserve (> 50,000 ha) in order to conserve species with large area requirements (large carnivores, elevational migrants).

3. For smaller reserves, how small is too small? Studies of nonflying mammals and edge effects in north Queensland rainforest suggest that reserves of less than 2,000–4,000 ha in area are unlikely to preserve representative samples of primary forest over the long term (Laurance 1990a, 1991a,b, 1994).

4. Habitats outside nature reserves will almost always be crucial for the persistence of some species, populations, and ecological processes (Franklin 1993; Lidicker, Chap. 1). Land managers must not restrict their attention to parks and corridors but must also work to conserve and rehabilitate key elements in the surrounding landscape.

Summary

In tropical Queensland, Australia, extensive studies during the past decade have yielded new insights into the responses of nonflying mammals to rainforest fragmentation. This research has been conducted on the Atherton Tableland, a key center of mammalian endemism that has experienced progressive clearing and fragmentation of the forest cover for more than 60 years.

In general, mammals and their habitats are strongly influenced by processes originating outside fragment boundaries. First, the matrix of modified habitats surrounding fragments acts as a selective filter: species that tolerate or exploit the matrix often remain stable or increase in fragments, whereas those that avoid the matrix decline or disappear. Second, edge effects and wind disturbance greatly influence the physiognomy, floristic composition, and microclimate of fragments, and these ecological changes in turn have a substantial impact on fragment mammal communities.

Arboreal mammals are particularly dependent on vegetation for movement and clearly were affected by the nature of modified habitats surrounding fragments. Fragments linked to "mainland" forest tracts by well-developed corridors of regrowth forest were richer in arboreal species than were similarly sized fragments that lacked such connections. Recommendations for the design of faunal corridors and nature reserve systems in the tropics logically follow from these and other considerations.

Acknowledgments

I thank W. Z. Lidicker Jr. for inviting me to participate in the ITC symposium on landscape ecology. O. O. Osunkoya, P. Green, and W. Z. Lidicker Jr. reviewed an earlier draft of the manuscript. Fieldwork was supported by the U.S. National Science Foundation, Wildlife Conservation International, World Wildlife Fund-U.S., World Wildlife Fund-Australia, M. A. Ingram Trust, Museum of Vertebrate Zoology, Sigma Xi, and Qantas Airlines. Current support is provided by a senior research fellowship awarded by the Wet Tropics Management Authority, Cairns, Queensland.

Literature Cited

Ambuel, B., and S. A. Temple. 1983. Area-dependent changes in the bird communities and vegetation of southern Wisconsin forests. *Ecology* 64:1057–68.

Bennett, A. F. 1990a. Habitat corridors and the conservation of small mammals in a fragmented forest environment. *Landscape Ecol.* 4:109–22.

Bennett, A. F. 1990b. *Habitat Corridors: Their Role in Wildlife Management and Conservation.* Melbourne, Victoria, Australia: Department of Conservation and Environment. 37 pp.

Brown, J. H., and A. Kodric-Brown. 1977. Turnover rates in insular biogeography: Effect of immigration on extinction. *Ecology* 58:445–9.

Burkey, T. V. 1989. Stochastic extinction in nature reserves: The effect of fragmentation, and the importance of migration between reserve fragments. *Oikos* 55:75–81.

Covacevich, J., and M. Archer. 1970. The distribution of the cane toad, *Bufo marinus,* in Australia and its effects on indigenous vertebrates. *Mem. Queensl. Mus.* 17:305–10.

Crome, F. H. J., J. Isaacs, and L. A. Moore. 1995. The utility to wildlife of remnant riparian vegetation and associated windbreaks in the tropical Queensland uplands. *Pac. Conserv. Biol.* 2. In press.

Diamond, J. M. 1984. "Normal" extinctions of isolated populations. Pages 191–246 *in* Nitecki, M. H. (ed.), *Extinctions.* Chicago: University of Chicago Press. 354 pp.

Diamond, J. M., K. D. Bishop, and S. Van Balen. 1987. Bird survival in an isolated Javan woodlot: Island or mirror? *Conserv. Biol.* 1:132–42.

Erwin, T. L. 1986. The tropical forest canopy. Pages 123–9 *in* Wilson, E. O., and F. M. Peter (eds.), *Biodiversity.* Washington, D.C.: National Academy Press. 521 pp.

Fahrig, L., and G. Merriam. 1985. Habitat patch connectivity and population survival. *Ecology* 66:1762–8.

Foster, R. B. 1980. Heterogeneity and disturbance in tropical vegetation. Pages 75–92 *in* Soulé, M. E., and B. A. Wilcox (eds.), *Conservation Biology: An Evolutionary-Ecological Perspective.* Sunderland, Mass.: Sinauer Associates. 395 pp.

Franklin, J. F. 1993. Preserving biodiversity: Species, ecosystems, or landscapes? *Ecol. Appl.* 3:202–5.

Frawley, K. J. 1983. *A History of Forest and Land Management in Queensland, with Particular Reference to the North Queensland Rainforest.* Report to the Rainforest Conservation Society of Queensland, Brisbane. 458 pp.

Gilpin, M. E., and J. M. Diamond. 1982. Factors contributing to non-randomness in species co-occurrences on islands. *Oecologia* 52:75–82.

Grant, P. R. 1966. Ecological compatibility of bird species on islands. *Am. Nat.* 100:451–62.

Harris, L. D. 1984. *The Fragmented Forest.* Chicago: University of Chicago Press. 211 pp.

62   WILLIAM F. LAURANCE

Henein, K., and G. Merriam. 1990. The elements of connectivity where corridor quality is variable. *Landscape Ecol.* 4:157–70.

Hobbs, R. J. 1993a. Effects of landscape fragmentation on ecosystem processes in the western Australian wheatbelt. *Biol. Conserv.* 64:193–201.

Hobbs, R. J. 1993b. Can revegetation assist in the conservation of biodiversity in agricultural areas? *Pac. Conserv. Biol.* 1:29–38.

Hopkins, M. S., J. Ash, A. W. Graham, J. Head, and R. K. Hewett. 1993. Charcoal evidence of the spatial extent of the *Eucalyptus* woodland expansions and rain forest contractions in North Queensland during the late Pleistocene. *J. Biogeogr.* 20:357–72.

Hughes, R. 1987. *The Fatal Shore: The History of the Colonization of Australia.* New York: Pergamon Press. 491 pp.

Janzen, D. H. 1983. No park is an island: Increase in interference from outside as park size decreases. *Oikos* 41:402–10.

Kapos, V. 1989. Effects of isolation on the water status of forest patches in the Brazilian Amazon. *J. Trop. Ecol.* 5:173–85.

Kershaw, A. P. 1986. Climatic change and Aboriginal burning in north-east Australia during the last two glacial-interglacial cycles. *Nature (London)* 322:47–9.

Keto, A., and K. Scott. 1986. *Tropical Rainforests of North Queensland: Their Conservation Significance.* Rep. 3. Canberra: Australian Heritage Commission. 195 pp.

Laurance, W. F. 1986. Logging and the survival of wildlife in Australia's tropical rainforests. *Heritage Aust.* Summer:5–9.

Laurance, W. F. 1987. The rainforest fragmentation project. *Liane* 25:9–12.

Laurance, W. F. 1989. *Ecological Impacts of Tropical Forest Fragmentation on Nonflying Mammals and Their Habitats.* Ph.D. thesis. Berkeley: University of California. 502 pp.

Laurance, W. F. 1990a. Comparative responses of five arboreal marsupials to tropical forest fragmentation. *J. Mammal.* 71:641–53.

Laurance, W. F. 1990b. Distributional records for two "relict" dasyurid marsupials in north Queensland rainforest. *Aust. Mammal.* 13:215–8.

Laurance, W. F. 1991a. Ecological correlates of extinction proneness in Australian tropical rain forest mammals. *Conserv. Biol.* 5:79–89.

Laurance, W. F. 1991b. Edge effects in tropical forest fragments: Application of a model for the design of nature reserves. *Biol. Conserv.* 57:205–19.

Laurance, W. F. 1992. Abundance estimates of small mammals in Australian tropical rainforest: A comparison of four trapping methods. *Wildl. Res.* 19:651–5.

Laurance, W. F. 1993a. The pre-European and present distributions of *Antechinus godmani* (Marsupialia: Dasyuridae), a restricted rainforest endemic. *Aust. Mammal.* 16:23–7.

Laurance, W. F. 1993b. Research challenges and opportunities in the Wet Tropics of Queensland World Heritage Area. *Pac. Conserv. Biol.* 1:3–6.

Laurance, W. F. 1994. Rainforest fragmentation and the structure of small mammal communities in tropical Queensland. *Biol. Conserv.* 69:23–32.

Laurance, W. F., J. Garesche, and C. W. Payne. 1993. Avian nest predation in modified and natural habitats in tropical Queensland: An experimental study. *Wildl. Res.* 20:711–23.

Laurance, W. F., and J. D. Grant. 1994. Photographic identification of ground-nest predators in Australian tropical rainforest. *Wildl. Res.* 21:241–248.

Laurance, W. F., and E. Yensen. 1991. Predicting the impacts of edge effects in fragmented habitats. *Biol. Conserv.* 55:77–92.

Leung, L. K. P., C. R. Dickman, and L. A. Moore. 1993. Genetic variation in fragmented populations of an Australian rainforest rodent, *Melomys cervinipes. Pac. Conserv. Biol.* 1:58–65.

Lovejoy, T. E., R. O. Bierregaard, A. B. Rylands, et al. 1986. Edge and other effects of isolation on Amazon forest fragments. Pages 257–85 *in* Soulé, M. E., and B. A. Wilcox (eds.), *Conservation Biology: The Science of Scarcity and Diversity.* Sunderland, Mass.: Sinauer Associates. 584 pp.

Osunkoya, O. O. 1994. Postdispersal survivorship of north Queensland rainforest seeds and fruits: Effects of forest, habitat and species. *Aust. J. Ecol.* 19:52–64.

Pahl, L. I., J. W. Winter, and G. Heinsohn. 1988. Variation in responses of arboreal marsupials to fragmentation of tropical rainforest in north-eastern Australia. *Biol. Conserv.* 46:71–82.

Saunders, D. A., and C. P. de Rebeira. 1991. Values of corridors to avian populations in a fragmented landscape. Pages 221–40 *in* Saunders, D. A., and R. J. Hobbs (eds.), *Nature Conservation 2: The Role of Corridors.* Chipping Norton, N.S.W., Australia: Surrey Beatty & Sons. 442 pp.

Saunders, D. A., R. J. Hobbs, and C. R. Margules. 1991. Biological consequences of ecosystem fragmentation: A review. *Conserv. Biol.* 5:18–32.

Simberloff, D., and J. Cox. 1987. Consequences and costs of conservation corridors. *Conserv. Biol.* 1:63–71.

Simberloff, D., J. A. Farr, J. Cox, and D. W. Mehlman. 1992. Movement corridors: Conservation bargains or poor investments? *Conserv. Biol.* 6:493–504.

Terborgh, J. 1975. Faunal equilibria and the design of nature reserves. Pages 369–80 *in* Golley, F., and E. Medina (eds.), *Tropical Ecological Systems: Trends in Terrestrial and Aquatic Research.* New York: Springer-Verlag. 585 pp.

Tracey, J. G. 1982. *The Vegetation of the Humid Tropical Region of North Queensland.* Indooroopilly, Queensland, Australia: CSIRO Long Pocket Labs. 124 pp.

Tracey, J. G. 1986. *Trees on the Atherton Tableland: Remnants, Regrowth and Opportunities for Planting.* Paper 1986/35. Canberra: Centre for Resource and Environmental Studies, Australian National University. 31 pp.

Van Dyck, S. 1985. *Sminthopsis leucopus* (Marsupialia: Dasyuridae) in north Queensland rainforest. *Aust. Mammal.* 8:53–60.

Webb, L. J., and J. G. Tracey. 1981. Australian rainforests: Patterns and change. Pages 605–94 *in* Keast, A. (ed.), *Ecological Biogeography of Australia.* The Hague: Junk. 2,061 pp.

Wet Tropics Management Authority (WTMA). 1992. *Wet Tropics Plan: Strategic Directions.* Cairns, Queensland, Australia: Wet Tropics Management Authority. 160 pp.

Wharburton, N. 1987. *The Application of the Theory of Island Biogeography to the Avifauna of Remnant Patches of Rainforest on the Atherton Tablelands with a View to Their Conservation.* M.Sc. thesis. Townsville, Queensland, Australia: James Cook University. 173 pp.

Wilcox, B. A. 1980. Insular ecology and conservation. Pages 95–117 *in* Soulé, M. E., and B. A. Wilcox (eds.), *Conservation Biology: An Evolutionary-Ecological Perspective.* Sunderland, Mass.: Sinauer Associates. 395 pp.

Willson, M. F., and F. H. J. Crome. 1989. Patterns of seed rain at the edge of a tropical Queensland rain forest. *J. Trop. Ecol.* 5:301–8.

Winter, J. W., F. C. Bell, L. I. Pahl, and R. G. Atherton. 1984. *The Specific Habitats of Selected Northeastern Australian Rainforest Mammals.* Report to World Wildlife Fund-Australia, Sydney. 212 pp.

Winter, J. W., F. C. Bell, L. I. Pahl, and R. G. Atherton. 1987. Rainforest clearfelling in northeastern Australia. *Proc. R. Soc. Queensl.* 98:41–57.

4

Movement in Spatially Divided Populations: Responses to Landscape Structure

Gray Merriam

Recently, attention has been drawn to the question of how mammals living in isolated habitat fragments move through the less usable parts of the mosaic surrounding their habitat fragment in order to access more usable habitat fragments and resource patches. As the study of movements came to focus on habitats left in a sea of nonhabitat by fragmentation, results required knowledge of the spatial structure of the landscape mosaic in addition to the classical focus on home ranges and territories. This development has given rise to alternative approaches to the study and interpretation of mammal movements across mosaic environments (now commonly called *landscapes*).

One interpretation of these new studies of movement across landscapes is simply that the unusual movement behavior observed results from the fragmentation of the habitat in a novel fashion and hence is peculiar to such new habitat patterns. If this is so, then it is important to learn about this movement behavior, and its relationship to population survival and evolution, because fragmentation and habitat patch isolation are characteristic of so much of the global habitat of mammals and are ever-increasing. In Chapter 6 of this volume, Bissonette and Broekhuizen discuss some effects of such extreme habitat changes on *Martes foina,* the stone marten.

This interpretation, however, begs the question of whether such movement among patches is absent in large, continuous habitats. The potential significance of this question has grown as we have recently accepted the ubiquity of heterogeneity (Kolasa and Pickett 1991) and its functional importance in ecosystem processes (Merriam 1991) and have learned how easily patchy patterns (and movements among them) can be masked if the wrong scale is selected when data are collected (Turner et al. 1989). Did the constrained movement patterns interpreted from earlier mammal studies arise from methods of data collection, such as undersized trap grids (Wegner and Merriam 1990)? Did mammals evolve the basis for interpatch movement behaviors long before fragmentation because continuous habitat is functionally patchy and always has provided selective forces for the evolution of such interpatch movement behaviors (Middleton and Merriam 1983; Southwood 1977)? Are these movements just easier to see now where habitats are spatially fragmented?

The nature of the operational demographic unit (ODU) is a fundamental problem that has not been stressed in classical mammalogy but that now must be confronted when a landscape ecological approach is used. The effects of movements on an ODU are clear from the roles of immigration and emigration and of births and deaths in any demographic change. Neither of these pairs of parameters is necessarily the controlling set; movements can control the dynamics of populations. When the ODU is a spatially divided population, it becomes critical to have an operational definition of the demographic unit; otherwise, the role of movements among patches (subpopulations of the ODU) can be misinterpreted demographically.

Spatially divided populations may often have indistinct spatial boundaries. Studying a small subunit of a spatially divided population may reveal only hopeless demographic stochasticity. In these instances, the spatial sample must be enlarged until the demographic processes are representative of both the determinism and the stochasticity of the ODU. In a successful spatially divided population, this point will be indicated by a negligible extinction rate for the population sample (Fahrig and Merriam 1994). If the spatial basis of demographics is erroneous or incomplete, then predictions of population survival will also be erroneous.

Linking Movement in Landscapes and Spatially Divided Populations

Movements by Mammals within Landscapes

Dispersion is used to represent the spatial scatter of individuals and their home sites, and *dispersal* is the general term for movements of those individuals among sites (Hansson 1991). *Landscape movement,* shorthand used by landscape ecologists, is synonymous with dispersal provided the spatial scale used is large enough to measure very long movements. All movements through the landscape mosaic, except continental migration, are included as well as all demographic classes. Analysis of landscape movements emphasizes possible constraints by *landscape composition* (types of resource patches) and *landscape configuration* (positioning of landscape elements) and by behavioral interactions with these two variables, which result in a third variable called *connectivity* (Bissonette and Broekhuizen, Chap. 6; Merriam 1984; Taylor et al. 1993).

Connectivity can be defined as the probability of movement of a species or a behavioral or demographic subgroup among landscape elements of a mosaic. To estimate this probability, landscape movement is measured, including any movements among landscape elements in the mosaic. Landscape movements are a function of the interactions between individual behavior and ecological landscape structure. Data on locations and extent of movements must be sensitive to the study species' behavior and not to the measurement methods. N. C. Stenseth and R. A. Ims (pers. comm.) have shown the effects of behavioral differences for even very similar subspecies. The magnitude of these movements may vary according to the scale of the patches chosen for study (i.e., the selected focal scale of heterogeneity) but may be recorded at a series of scales in the appropriate range.

Population Division and Evolutionary Selection

Spatial division of populations, whether enforced by natural heterogeneity of resources or by technocultural fragmentation of habitat, can act as a constraint on population success. The constraint is manifested as requirements for more frequent and/or more extensive landscape movements. At the limit, if landscape movement is inadequate, the spatially divided population may fail, causing a landscape-scale extinction.

Such realizations have led to assessment of the relative potential of responses to such population declines by the classically expected target of evolutionary selection—the reproductive capacity of individuals. Fahrig and Paloheimo (1988) showed in simulations that any of several variables affecting landscape movements (e.g., range of movement, frequency of leaving a patch) were significantly more effective in maintaining the population under such conditions than the classically proposed target for natural selection (individual reproduction). Earlier, Comins et al. (1980) had found a solution to an *evolutionarily stable strategy* in terms of a single variable related to movement, implying the same conclusion (Merriam 1984).

The realization that dispersal has adaptive value is not new (Hansson, Chap. 2; Laurance, Chap. 3), but the further realization that the constraint imposed by the landscape has the potential to overcome selection for individual reproductive characteristics is important to our views of differential species survival and of evolution. That modified movement behavior could be the most effective response to strong selective pressures leads to the prediction of adaptive responses to the selective forces of technoculturally fragmented habitats (Merriam 1991). It can be argued that the rate and intensity of changes in landscape patterns caused by human technical developments far exceed those of any sustained landscape changes experienced during the evolution of most mammals. Selection pressure may be increasing for rapid response to demographic threats caused by these rapid and powerful impacts, especially in extreme situations such as the urban environments discussed by Kozakiewicz and Szacki (Chap. 5) and Bissonette and Broekhuizen (Chap. 6).

Alternatively, if natural mosaics have been patchy historically, then landscape movement behavior of spatially divided populations could have evolved in those mosaics. If the patches in those mosaics were subject to intense and rapid change from natural disturbances, such as fires, then natural selection may always have fostered adaptive landscape movement behavior as well as adaptive individual reproductive characteristics.

Movements among Subpopulations and Population Success

The effects of movements among spatially divided subpopulations on the whole population involve a complex of variables. The simplest is *recolonization* after local (patch) extinctions (Fahrig and Merriam 1985, 1994; Henderson et al. 1985; Merriam 1984). Related, but less easy to measure empirically, is *supplementation* of small subpopulations that raises them above the numerical threshold below which stochastic effects overcome evolutionarily directed demography. The combination of colonization and supplementation can be critical to a spatially divided population (Fahrig and Merriam 1985, 1994; Hanski 1990). Either effect can simply increase mortality if the patches receiving the movers are death traps. However, if mortality is not immediate, then so-called *submarginal* or *sink* habitat patches can supply rescuers to local failures in other patches and thus perform a rescue function similar to that of any other patch in a spatially divided population (Brown and Kodric-Brown 1977; Henderson et al. 1985).

This rescue effect can be extensive where the several-fold expansion in habitat that is needed just after annual reproduction is not available in good-quality habitat but is available as "expansion patches" of submarginal habitat. Such seasonally adequate, submarginal habitat resources, in addition to the limited prime habitat areas, can accommodate juveniles until they are recruited into prime habitat. This is the case for most recruits into some farmland chipmunk populations that retract into prime patches before hibernation (Bennett et al. 1994; K. Henein and G. Merriam, unpub. data).

Interpatch movements may "filter" and change the age, sex, and behavioral structure of patch populations (Merriam et al. 1989; Szacki 1987). The genetic effects of interpatch movements (gene flow) should always exceed the demographic effects (recolonization, rescue) over the same interpatch distances, because the effective threshold of movements needed for genetic continuity is much lower than that for demographic effects, and the allowable time frame is much more permissive (Merriam et al. 1989).

Limiting Cases of Spatial Subdivision

Barriers limit the distribution of spatially divided populations because they impose an unusually "hard" edge on the spatially divided population. Such a well-defined boundary may not be present over the rest of the range of the spatially divided population. *Connectivity,* the inverse of functional isolation, implies connection of subpopulations by movements within an ODU. Spatial division of populations means that the ODU is patchy. If separations between these patches are taken as population boundaries and if patch populations are thus not seen as subpopulations, then results can be misinterpreted. If data are collected from only a single patch or too few patches, the sampled information will not support confident prediction about the ODU. Stochastic demographic behavior of individual patches will make prediction difficult or impossible unless the sample is enlarged to include enough patches or subpopulations to reduce the probability of extinction for the whole sample to a low value. Until this sample size is reached, stochastic variation will dominate the dynamics of the demographic unit and preclude deterministic prediction (Fahrig and Merriam 1994). Prediction can be done with either a statistical model (Gotelli and Kelly 1993; Lankester et al. 1991; Verboom et al. 1991) or an individual-based model (Henein and Merriam 1990; Judson 1994). An expanded sample of a spatially divided population that meets the criterion of deterministic predictability also is a good general representation of the ODU, that is, the population.

A clear geographic boundary for a spatially divided population would require a zone across which no interpatch movements take place. This is unusual except where an absolute barrier exists (Merriam et al. 1989). In much of the mammalogical literature, geographic boundaries for demographic units have been assumed based on habitat patch boundaries. For spatially divided populations, such an assumption could easily lead to defining a single-patch population as the ODU. This would be valid only if the study patch were big enough to escape stochastic, small-population effects or if the demographics did not include any deterministic effects. If the focus is at the level of the individual patch and not on measurements at the population level, such an erroneous definition of the population could easily result in misinterpretation of the demographic dynamics. This reasoning does not depend on any theory from equilibrium island biogeography. That theory does not include single-species demographic variables; it only predicts equilibrium numbers of species, and demographic studies based on it are subject to potentially serious errors.

The data needed to relate dynamics of a demographic unit to the spatial variables that may drive those dynamics must represent the demographically complete population or a sample that can represent the whole population. The only possible boundary of a spatially divided population is a zone with severely reduced or no interpatch movements. The spatial extent of an ODU encompasses interpatch zones with a gradient of rates of interpatch movements varying from the rate within a patch to zero between two unconnected patches. We can measure these rates and arrange the interpatch zones into isoflux zones. In reality, the gradient of flux rates usually does not extend to zero except where there is a strong barrier effect, such as a large river, a major habitat boundary, or a technocultural barrier (Merriam et al. 1989). A spatial plot of isoflux zones may be helpful in assessing population dynamics underlying biodiversity changes and in designing management units and interventions for spatially divided populations under the influence of manipulable forces of fragmentation (Merriam 1991).

The distance scale over which the gradient of flux rates ranges from the within-patch rate to near zero at the absolute boundary of the population will be a function of the landscape movement range of the members of the population and will determine the scale of the ecological neighborhood (Addicott et al. 1987). These scales are not necessarily a function of the movement distances cited in classical mammalogy studies done within single habitat units. Landscape movement ranges may be an order of magnitude greater than movements within patches (Henderson et al. 1985; K. Henein and G. Merriam, unpub. data; Kozakiewicz and Szacki, Chap. 5; Wegner and Merriam 1990, unpub. data). Measuring movements over the annual landscape range of an individual may not be possible with standard methods, even for small mammals. Not only does livetrapping inevitably truncate individual movements, but also the labor requirements of humane livetrapping commonly restrict sampling to spatial scales that are too small to measure accurately the parameters needed for individual-based models of spatially divided populations.

The requirements for identifying the ecological neighborhood and its scale can be related to landscape parameters often called "the three Cs": composition, configuration, and connectivity. The first need is a measure of the composition of resource patches used; for this, identity of individuals is less important than spatial scale, timing, and confidence limits of landscape movement data. Enlarging the scale of the sampling with methods such as track-registry stations (Merriam and Wegner 1992) or color-coded feces (Merriam et al. 1989; Szacki and Liro 1991) produces better measures for the study of composition.

To relate landscape movement data to landscape configuration and connectivity, movement flux data should include records of all movements between habitat patches, including unpredictable movements, at scales determined by large landscape structures, such as major barriers or other features that may impose zones of equal probability of movement. For these records, data on individual identity, explicit spatial origin, and possibly age, sex, or behavioral class are needed. Telemetric tracking over appropriate spatial scales is significantly better than trapping, even for small mammals, because the spatial scale is less constrained and predetermined, and because movements are not truncated with each datum. Nest boxes, used to ear-tag and recapture, are equally good; they can be arrayed in large numbers spread over areas too large to attend daily and can provide data on natality, genetic relationships, social structuring, dispersal groups, food caching, and other variables (Goundie and Vessey 1986; Morris 1986, 1989; Nicholson 1941; J. Wegner and G. Merriam, unpub. data).

The suitability of data for assessing connectivity is controlled by the large spatial scale required for spatially divided populations of medium-sized to large mammals. The scale limits availability of connectivity data by reducing sample size and the completeness of telemetric searching. The appropriate spatial scale for divided populations of medium-sized to large mammals is poorly known. Landscape movements of individuals must be synthesized into ODUs if demographic analysis is to be related to landscape pattern and its changes. Impacts of technocultural activities, such as forest harvesting or expansion of transportation infrastructure, need to be evaluated in terms of their potential for degrading success of populations, especially spatially divided populations. Timber wolves of Canada's Banff National Park move almost as far as Yellowstone National Park (P. C. Paquet, pers. comm.; Purves et al. 1992) and, during repeated recolonizations, easily could have connected through the Jasper Park population to the Yukon. The routes available for these connections among spatially divided subpopulations are difficult to monitor in the Rocky Mountains, and data are subject to potential "highway bias." Rates of movements over such long distances are equally difficult to measure confidently.

Linking Movements and Landscape Structure

Even continuous habitat usually is not homogeneous but rather is a mosaic of patches. A *mosaic* is a differential distribution of resources that produces gradients of resource availability that cause flows of resources or of the organisms using them. The concept of connectivity also applies in such continuous mosaics, although empirical evidence on movements in habitats such as continuous forest is not commonly formulated in terms of connectivity. There is some evidence that some species do have adaptations for landscape-scale movements even when their evolutionary history has primarily been in forests (Middleton and Merriam 1983). There also is some evidence that species populations may use one section of a continuous habitat mosaic and not use a contiguous section that is apparently indistinguishable ecologically (Krohne and Burgin 1990; Middleton and Merriam 1983). The null model for landscape movements in continuous habitat mosaics should be the same as for fragmented mosaics.

Fragmented mosaics vary from nearly continuous, with only a minor part of the area modified by habitat removal, to the opposite, with only isolated fragments of the original mosaic remaining in a matrix of introduced alien habitats, such as woodlots in farmland. Changing minor amounts of habitat affects landscape composition only if it removes or adds some particularly vital resource (Dunning et al. 1992; Taylor et al. 1993). Minor habitat changes are likely to affect connectivity in a mosaic only if the habitat removed is strategically located spatially. For example, removal of the only mineral lick in an intermontane valley has both a composition effect and a configuration effect, with a final interaction effect on connectivity. Insertion of a linear barrier, such as a road, can be analogous to removal of a resource; addition of a limiting resource will have the opposite effect on the same variables. Except where blocked by barriers, movements would be expected to be both qualitatively and quantitatively similar to those in continuous mosaics until fragmentation has affected more than a minor part (and possibly, a major part) of the favored habitat in the mosaic (L. Fahrig, unpub. data).

Intermediate to severe fragmentation (the range between removal of the majority of the favored habitat from the mosaic and removal of all but isolated fragments) creates a critical stage for the survival of the indigenous species of the mosaic. The severe end of this gradient is the current condition in many areas of industrialized farmland. In these circumstances, managers and activists may simplify the concept of connectivity and advance "corridors" as a putative, overriding remedy. However, theoretical and simulation studies have demonstrated that it is *connectivity,* not corridors, that is vital to spatially divided populations (Fahrig and Merriam 1985; Hanski 1990). Corridors offer no special gains in connectivity and are, by definition, less beneficial to populations than habitat restoration would be.

Many studies have related the occurrence of various mammals to corridors of various types (see Laurance, Chap. 3). Others have directly measured use of corridors by various mammals either for movements that can be related to connectivity (Merriam and Lanoue 1990) or for both movement and habitat (Bennett et al. 1994; Laurance, Chap. 3). However, neither connectivity nor corridors have been shown in empirical experiments to be essential to the survival of spatially divided populations in a way that satisfies the strictest scientific critics. Management trials based on models such as that for conservation of the badger in The Netherlands (Lankester et al. 1991) may provide nonexperimental evidence on this point. With clear predictions from models, uncontrolled experiments, such as the nearly complete isolation of a California cougar population by housing developments, may provide tests of the terminal role of corridors for chronically degraded populations (Beier 1993).

The advocacy of corridors is most common in situations where only isolated fragments of habitat remain, and in these circumstances corridors can be expected to increase landscape movement. Reliance on selection pressure to increase individual reproductive contributions is unlikely to work, as discussed above. However, corridors that will facilitate landscape movement must be selected, or designed and placed, so that they are behaviorally acceptable for movement by the target species. Land planning exercises commonly use mappable vegetation as the sole criterion for corridor utility, and the telemetry and behavioral studies that are needed to design and place the corridors appropriately often are omitted (Hudson 1991; Woods 1990). When a large habitat unit is added as a connector to other large units, the connection provides for both habitat and movement (Harris and Gallagher 1989). However, where point crossings of highways or other barriers are required or where narrow, linear vegetation elements are expected to act as connectors, composition and placement are important (Beier 1993; Bennett et al. 1994). If, as is often the case, populations have already been severely degraded in performance, corridors may serve only to "buy time." In these situations, structure and placement are important because the degraded populations may not have time for behavioral accommodation to poor designs.

Because corridors are a last-ditch substitute for providing more comprehensive connectivity by modifying landscape composition and configuration, the chance that they might increase connectivity and prolong survival of a valued but degraded population should be taken if the "precautionary principle" is followed (Hobbs 1992). However, to advocate corridors as a final solution is foolish; the landscape conditions that put the population in jeopardy must be corrected.

Critics of conservation and movement corridors have cited several ecological risks (Hobbs 1992; Simberloff and Cox 1987; Simberloff et al. 1992). In general, however, the available support for the reality of these risks is as weak as that for the proposition that corridors enhance the success of spatially divided populations. An apparent exception is the contention that corridors increase movement of weeds, but the appropriate test of this hypothesis is to show that the matrix of a fragmented landscape does not promote weed movement more than corridors do, and this demonstration is lacking.

The discussion of corridors as a measure for conservation quickly loses its scientific nature and enters the policy arena when "costs," "bargains," and "investments" enter the arguments. The interesting scientific questions are about connectivity, its evolutionary role in population theory, and what adaptations various species make when their connectivity is affected by landscape modification, not just in severely isolated fragments but over the entire gradient of fragmentation and spatial scales.

Movements and Survival in Future Landscapes

The greatest threat to spatially divided mammal populations is that fragmentation will progress to the point where most of the original mosaic has been removed, only fragments remain, and connectivity is severely limited. Under these conditions, as shown by Fahrig and Paloheimo (1988), the effects of rapid and severe fragmentation cannot be moderated by changes in individual demographic characteristics as effectively as they can by changes in individual landscape movement behavior.

In the future, populations in increasingly fragmented habitat should survive best if their behavior is sufficiently flexible to allow increased landscape movement as a temporary response to rapid and intense fragmentation. Evolutionary selection should subsequently favor individuals and species that show such behavioral flexibility and expanded movement capability. Success should be promoted both by demographic effects and by increased ability to access and use complementary resource patches. Also favored will be those able to gain supplementation from novel resource patches that accompany fragmentation changes (Dunning et al. 1992; Taylor et al. 1993). Species that lack flexibility in movement behavior and in resource selection should be expected to degrade demographically and to decline, with an imbalance of local extinctions over recolonizations that may eventually result in regional and larger-scale extinctions (Ehrlich and Daily 1993; Merriam and Wegner 1992).

Such predictions are emerging from a study by K. Henein and G. Merriam (unpub. data), which compares simulations of *Peromyscus leucopus* and *Tamias striatus* populations in a spatially explicit geographic information system (GIS) simulation of a common landscape. The models are parameterized from empirical studies by John Wegner and by Kringen Henein. Both species expand their movement capabilities, but *P. leucopus* moves through all parts of the landscape except grass-dominated hay or pasture, whereas *T. striatus* is constrained to move only in wooded habitat and fencerows. It is also limited to these habitats for all of its resource use, while *P. leucopus* exhibits both complementation (moving among and using resources from several patch types) and supplementation (use of farm crops and buildings). When increased fragmentation is programmed into the GIS spatial model of the landscape, the decided advantage of *P. leucopus* over *T. striatus* under probable future conditions is shown. Such studies may give us clues to mammalian characteristics that will predict species survival in future landscapes.

Recommendations

The results of new studies of mammal populations and faunal groups should be presented with an explicit spatial basis; very few situations permit meaningful interpretation of their dynamics without the accompanying spatial relationships. Landscape ecology provides a framework for this approach.

Before initiating a population study, researchers should determine a neighborhood scale as well as scales for each specific question by sampling over nested scales. Because results reported in the literature commonly were obtained at a single, predetermined scale, they are not an appropriate guide to scale selection. Neighborhood and movement scales are, by definition, variable according to the composition and configuration of the landscape.

Because the four basic variables in population dynamics can determine one steady state derived from movements (immigration minus emigration) and one steady state derived from reproduction and mortality (births minus deaths), it is critical to distinguish between these pairs of variables in the field. As movement range increases, distinguishing deaths from emigration for any population unit becomes more difficult or, if the scale of study is too small, impossible. Because moving individuals usually are more detectable than dead ones, measuring movements should have priority. Livetrapping has such clear bias that it should not be used for measuring movements without calibration by alternative methods.

Restructuring of the landscape and of faunal assemblages is creating many novel relationships for mammal populations. Therefore, reliance on classical sampling methods may not be sufficient. Neither distribution of sampling effort (resource types, time and space scales) nor measuring or indexing methods should omit any possibilities simply because they may be unconventional.

Summary

An objective of many studies of mammals is to gain predictive understanding of population changes. Meeting this objective requires data for a population sample that is demonstrably an ODU. The demographic unit may be spatially divided either by natural heterogeneity of habitats or by technocultural fragmentation of habitat. In either case, movement of individuals through the habitat mosaic, or landscape, interconnects subpopulations demographically, and the extent and frequency of these movements define the ODU.

In addition, these movements can control the demographic success of the population more effectively than can variation in reproductive characteristics. The effects of movements on population success depend on habitat conditions, which can be measured by composition and configuration of the landscape, and their interactions with movement behavior. The results of this interaction determine landscape connectivity. Predictions of the survival of mammal populations should be based on relationships between demographics and the structure of and movements through landscape mosaics.

Acknowledgments

Brett Goodwin, John Pedlar, and John Wegner suggested significant improvements to earlier versions of the manuscript.

Literature Cited

Addicott, J. F., J. M. Aho, M. F. Antolin, D. K. Padilla, J. S. Richardson, and D. A. Soluk. 1987. Ecological neighbourhoods: Scaling environmental patterns. *Oikos* 49:340–6.
Beier, P. 1993. Determining minimum habitat areas and habitat corridors for cougars. *Conserv. Biol.* 7:94–108.
Bennett, A., K. Henein, and G. Merriam. 1994. Corridor use and the elements of corridor quality. *Biol. Conserv.* 68:155–65.

Brown, J. H., and A. Kodric-Brown. 1977. Turnover rates in insular biogeography: Effect of immigration on extinction. *Ecology* 58:445–9.

Comins, H. N., W. D. Hamilton, and R. M. May. 1980. Evolutionarily stable dispersal strategies. *J. Theor. Biol.* 82:205–30.

Dunning, J. B., B. J. Danielson, and H. R. Pulliam. 1992. Ecological processes that affect populations in complex landscapes. *Oikos* 65:169–75.

Ehrlich, P. R., and G. C. Daily. 1993. Population extinction and saving biodiversity. *Ambio* 22:64–8.

Fahrig, L., and G. Merriam. 1985. Habitat patch connectivity and population survival. *Ecology* 66:1762–8.

Fahrig, L., and G. Merriam. 1994. Conservation of fragmented populations. *Conserv. Biol.* 8:50–9.

Fahrig, L., and J. Paloheimo. 1988. Determinants of local population size in patchy habitats. *Theor. Popul. Biol.* 34:194–213.

Gotelli, N. J., and W. G. Kelly. 1993. A general model of metapopulation dynamics. *Oikos* 68:36–44.

Goundie, T. R., and S. H. Vessey. 1986. Survival and dispersal of young white-footed mice born in nest boxes. *J. Mammal.* 67:53–60.

Hanski, I. 1990. Single-species metapopulation dynamics: Concepts, models and observations. *Biol. J. Linn. Soc.* 42:17–38.

Hansson, L. 1991. Dispersal and connectivity in metapopulations. *Biol. J. Linn. Soc.* 42:89–103.

Harris, L. D., and P. B. Gallagher. 1989. New initiatives for wildlife conservation: The need for movement corridors. Pages 11–34 *in* Mackintosh, G. (ed.), *In Defense of Wildlife: Preserving Communities and Corridors*. Washington, D.C.: Defenders of Wildlife. 96 pp.

Henderson, M. T., G. Merriam, and J. Wegner. 1985. Patchy environments and species survival: Chipmunks in an agricultural mosaic. *Biol. Conserv.* 31:95–105.

Henein, K., and G. Merriam. 1990. The elements of connectivity where corridor quality is variable. *Landscape Ecol.* 4:157–70.

Hobbs, R. J. 1992. The role of corridors in conservation: Solution or bandwagon? *Trends Ecol. Evol.* 7:389–92.

Hudson, W. E. (ed.). 1991. *Landscape Linkages and Biodiversity*. Washington, D.C.: Island Press. 196 pp.

Judson, O. P. 1994. The rise of the individual-based model in ecology. *Trends Ecol. Evol.* 9:9–14.

Kolasa, J., and S. Pickett (eds.). 1991. *Ecological Heterogeneity*. Ecological Studies 86. New York: Springer-Verlag. 332 pp.

Krohne, D. T., and A. B. Burgin. 1990. The scale of demographic heterogeneity in a population of *Peromyscus leucopus. Oecologia* 82:97–101.

Lankester, K., R. C. van Apeldoorn, E. Meelis, and J. Verboom. 1991. Management perspectives for populations of the European badger (*Meles meles*) in a fragmented landscape. *J. Appl. Ecol.* 28:561–73.

Merriam, H. G. 1984. Connectivity: A fundamental ecological characteristic of landscape pattern. Pages 5–15 *in Proceedings of the First International Seminar on Methodology in Landscape Ecological Research and Planning*. Roskilde, Denmark: International Association for Landscape Ecology. 118 pp.

Merriam, G. 1991. Corridors and connectivity: Animal populations in heterogeneous environments. Pages 133–42 *in* Saunders, D. A., and R. Hobbs (eds.), *Nature Conservation 2: The Role of Corridors*. Chipping Norton, N.S.W., Australia: Surrey Beatty & Sons. 442 pp.

Merriam, G., M. Kozakiewicz, E. Tsuchiya, and K. Hawley. 1989. Barriers as boundaries for metapopulations and demes. *Landscape Ecol.* 2:227–35.

Merriam, G., and A. Lanoue. 1990. Corridor use by small mammals: Field measurement for three experimental types of *Peromyscus leucopus. Landscape Ecol.* 4:123–31.

Merriam, G., and J. Wegner. 1992. Local extinctions, habitat fragmentation, and ecotones. Pages 150–69 *in* Hansen, A. J., and F. di Castri (eds.), *Landscape Boundaries: Consequences for Biotic Diversity and Ecological Flows.* Ecological Studies 92. New York: Springer-Verlag. 452 pp.

Middleton, J., and G. Merriam. 1983. Distribution of woodland species in farmland woods. *J. Appl. Ecol.* 20:625–44.

Morris, D. W. 1986. Proximate and ultimate controls on life history variation: The evolution of litter size in white-footed mice (*Peromyscus leucopus*). *Evolution* 40:169–81.

Morris, D. W. 1989. Density-dependent habitat selection: Testing the theory with fitness data. *Evol. Ecol.* 3:80–94.

Nicholson, A. J. 1941. The home and social habitats of the wood-mouse *Peromyscus leucopus noveboracensis* in southern Michigan. *Am. Midl. Nat.* 25:196–223.

Purves, H. D., C. A. White, and P. C. Paquet. 1992. *Wolf and Grizzly Bear Habitat Use and Displacement by Human Use in Banff, Yoho, and Kootenay National Parks: A Preliminary Analysis.* Banff, Alberta: Canadian Parks Service. 41 pp.

Simberloff, D., and J. Cox. 1987. Consequences and costs of conservation corridors. *Conserv. Biol.* 1:63–71.

Simberloff, D., J. A. Farr, J. Cox, and D. W. Mehlman. 1992. Movement corridors: Conservation bargains or poor investments? *Conserv. Biol.* 6:493–504.

Southwood, T. R. E. 1977. Habitat, the templet for ecological strategies? *J. Anim. Ecol.* 46:337–65.

Szacki, J. 1987. Ecological corridors as a factor in determining structure and organization of bank vole populations. *Acta Theriol.* 32:31–44.

Szacki, J., and A. Liro. 1991. Movements of small mammals in the heterogeneous landscape. *Landscape Ecol.* 5:219–24.

Taylor, P. D., L. Fahrig, K. Henein, and G. Merriam. 1993. Connectivity is a vital element of landscape structure. *Oikos* 68:571–3.

Turner, M. G., V. H. Dale, and R. H. Gardner. 1989. Predicting across scales: Theory development and testing. *Landscape Ecol.* 3:245–52.

Verboom, J., K. Lankester, and J. Metz. 1991. Linking local and regional dynamics in stochastic metapopulations. *Biol. J. Linn. Soc.* 42:39–55.

Wegner, J., and G. Merriam. 1990. Use of spatial elements in a farmland mosaic by a woodland rodent. *Biol. Conserv.* 54:263–76.

Woods, J. G. 1990. *Effectiveness of Fences and Underpasses on the Trans-Canada Highway and Their Impact on Ungulate Populations Project.* Calgary, Alberta: Canadian Parks Service, Western Region. 57 pp.

5

Movements of Small Mammals in a Landscape: Patch Restriction or Nomadism?

Michał Kozakiewicz and Jakub Szacki

For decades, few population ecologists took into account problems of landscape heterogeneity, seemingly not recognizing their possible influence on population dynamics. At present, just the opposite is true: the role of heterogeneity is stressed in almost every paper on population ecology of small mammals, even if it is not actually needed. Apparently, the problems of heterogeneity have become a priority for many scientific institutions. This does not mean, however, that heterogeneity is a clear and explicit concept.

First, there is a problem of scale, which is well illustrated by the example of a block of Swiss cheese (Milne 1991): its mass is quite different when estimated from within an air bubble than when measured as an average for the whole block. Problems related to scale are appreciated in landscape ecology mainly in theoretical papers (Kolasa and Rollo 1991; Lord and Norton 1990; Merriam 1988, 1990; Wiens 1976, 1985) and more specifically for small mammals (e.g., Stenseth and Lidicker 1992). Empirical population studies at different scales are very rare; a paper by Krohne and Burgin (1990) is one of a very few.

Second, the degree of heterogeneity depends on the size and life history of species being considered. Clearly, no systems are absolutely homogeneous, but a given habitat may be almost homogeneous for an insect, heterogeneous for a rodent, and part of a heterogeneous territory of a bird of prey (Begon et al. 1990).

Third, landscape heterogeneity may be seen either as patchiness, determined by the size and number of individual "grains" of habitat ("fine- and coarse-grained landscapes"), or as a range of habitat types occurring within a landscape. The first aspect of patchiness is described by landscape physiognomy, the second by landscape composition (Dunning et al. 1992).

Yet another factor complicating the problem of heterogeneity is time (Southwood 1977; Wiens et al. 1986). The role of particular patches in animal life may change over time, and animal behavior may be interpreted in different ways depending on the time scale (Lidicker and Stenseth 1992).

Researchers considering spatial behavior of animals in complex landscapes should bear in mind all these aspects of heterogeneity. In fact, most studies dealing with animal spatial ecology do consider problems of heterogeneity. We mean here, for instance, papers on habitat selection (e.g., Morris 1992), dispersal and its role in population regulation (Fahrig 1990; Gliwicz 1989, 1992; Lidicker 1975, 1988), and evolutionary processes in heterogeneous environments (Brown and Pavlovic 1992; Lidicker and Patton 1987).

In this chapter, we present possible relations between long-distance movements of small mammals and the landscape structure; that is, we attempt to answer the question of how landscape heterogeneity influences space use patterns of small mammals and, as a consequence, social and spatial population structure. We also discuss the implications of animal adaptations to heterogeneous landscapes for conservation. Because almost all authors have assumed a rather restricted range of small-mammal movements, that aspect of heterogeneity is not very prevalent in the scientific literature (but see Merriam, Chap. 4).

What Do We Actually Know about Small-Mammal Movements?

Over the last decades, a great body of information on small-mammal movements has been published. Most of this literature is devoted mainly to either dispersal or spatial activity within home ranges. The papers on dispersal deal primarily with its possible causes and population consequences; actual movement distances of dispersing individuals were not studied in detail. Data on movements within home ranges are not consistent among studies because the results strongly depend on methodology (number of captures used as the basis for calculations, assumed shape of home ranges, grid size, period of trapping, kind of bait used, etc.); see, for example, Andrzejewski and Babińska-Werka (1986) and Kozakiewicz et al. (1994).

Usually, livetrapping is used in studies on small-mammal movements, and it seems to be an inadequate sampling method. The presence of live traps certainly changes the spatial behavior of animals, because they spend a lot of time in traps eating artificially supplied food (Sheppe 1967). Bock and Babińska (1967), Buchalczyk and Pucek (1968), and many others have studied and discussed the influence of type of bait on trapping results. These studies showed that more animals are caught in baited traps than in unbaited ones partly because the animals remember locations of good meals (bait stations) for at least two weeks.

Another problem is that because trapping grids usually cover only a few hectares or less, long-distance movements cannot be detected (Szacki et al. 1993) or at least are less likely to be recorded than movements over short distances (Stenseth and Lidicker 1992). As a consequence of this bias, ecologists working with small mammals have assumed for many years that small home ranges (usually less than 1 ha) were typical. Those few data indicating that small mammals could move considerable distances were either disregarded or treated as an extraordinary event ("sallies outside the home range") of no importance for the theory of populations. Now that landscape ecology is gaining in importance and because landscape studies encourage investigations on larger spatial scales, an increasing amount of evidence has accumulated indicating that movements at distances of several hundred meters or even a few kilometers are far from exceptional (Table 5.1). We believe that long-distance movements, whatever their nature, play an important role in the life of small mammals and may significantly influence population structure and regulation.

Movements in a Landscape

It seems plausible that the greater the fragmentation of a landscape, the longer the distance of small-mammal movements. This correlation has been appreciated by several authors in different contexts. Brown and Pavlovic (1992) claim that movements between habitats are frequent under fine-grained heterogeneity but need not occur in coarse-grained landscapes. Harrison (1991) suggests that high interpatch dispersal is common among species living in unstable, transient, and/or patchy habitats.

Unfortunately, there is no strong evidence for increased animal mobility in heterogeneous landscapes. Population studies on the scale of a few kilometers are rare, and studies comparing homogeneous and heterogeneous landscapes in this regard are almost nonexistent. We are aware of only two studies that used the same method in both heterogeneous and homogeneous habitats and showed that heterogeneity does increase mobility of small mammals (J. Babińska-Werka, pers. comm.; Kozakiewicz et al. 1993). Kozakiewicz et al. (1993) used marked bait to follow the movements of bank voles (*Clethrionomys glareolus*) in homogeneous and heterogeneous environments. The mean distance between bait station and trapping point was almost twice as long in heterogeneous habitat (243 m, standard error 133 m) as in homogeneous habitat (135 m, standard error 116 m). The difference was statistically

Table 5.1. Maximum distances moved by small mammals as reported in various studies

Species	Habitat	Maximum distance (m)	Authors
Reithrodontomys megalotis	Heterogeneous prairie	3,200	Clark et al. 1988
Apodemus sylvaticus	Farmland	4,000	Tew 1988
A. sylvaticus	Woodland	1,030 (M), 810 (F)	Wolton 1985
Clethrionomys glareolus, A. sylvaticus	Urban mosaic	500	Dickman and Doncaster 1989
A. sylvaticus	Woods or fields	1,050	Bauchau and LeBoulengé 1991
Peromyscus leucopus	Agricultural mosaic	>1,000	Wegner and Merriam 1990
A. agrarius, C. glareolus	Suburban mosaic	>1,500	Szacki and Liro 1991
Microtus arvalis	Agricultural mosaic	500	von Somsook and Steiner 1991
Thomomys bottae	Grassland or forest or riparian habitat	>700	Lidicker and Patton 1987
Microtus xanthognathus	Burned or unburned forest	800	Lidicker and Patton 1987
Mus musculus	Welsh island	1,500	Lidicker and Patton 1987

significant (Student's t-test, $p < 0.05$). However, the maximum distances covered by voles were close to 500 m in both habitats (440 and 480 m in homogeneous and heterogeneous habitats, respectively) (Kozakiewicz et al. 1993).

In fact, most data on long-distance movements of small mammals have been obtained in heterogeneous landscapes (Table 5.1). That in itself is indirect evidence of the role of habitat heterogeneity. This correlation may be an artifact, however, because the larger the scale of observation, the greater the chance of finding heterogeneous habitats.

In any case, evidence is increasing that small mammals can and actually do move long distances. Long-distance movements must be difficult and dangerous for such small creatures and also very expensive in terms of individual energetics. Thus, individuals must have very good reasons to move, and we believe that heterogeneous environments provide such reasons. The greater the fragmentation (that is, the smaller the grains of habitats), the higher the chance that a single patch will contain only one type of necessary resource.

Resources in a single patch may easily run out, forcing animals to find new supplies in another patch. Thus, individuals may have to switch from one patch to another to find what they need (food of different types, shelters, mates, etc.) (Kozakiewicz 1995). Moreover, heterogeneity may inherently stimulate mobility, because there is always something new to find in different patches (landscape complementation and landscape supplementation, according to Dunning et al. 1992). In fine-grained landscapes, different patches of habitat are close to each other, so animals can easily explore several habitats and may reap benefits without much effort.

Variability of a landscape in time also encourages animals to move. A suitable habitat may become inhospitable after a time, and then animals have to move to another location. Andrzejewski (1963) described dramatic seasonal changes in the suitability of some parts of the Kampinos Forest (central Poland) for bank voles. Voles prefer Salici-Franguletum and Circaeo-Alnetum forest types, and old adults—that is, the socially dominant individuals in the population—occupy in autumn mainly the preferred habitats. However, in Kampinos Forest these habitats are flooded in late winter by rising groundwater, and dominants are forced to leave the sites they occupy. Winter dispersal, therefore, is the only way to survive for those individuals that in autumn established their residency in the best-quality sites in the forest (Andrzejewski 1963).

Moreover, requirements may change over time or seasons; for instance, animals may need different types of resources, and consequently different patches, during the breeding season than at other times. Kozakiewicz and Gortat (1994) studied seasonal dynamics of the spatial distribution of bank voles in Polish woodlots varying in area and degree of isolation within patchy agricultural landscapes. The seasonal dynamics of vole numbers differed markedly in different woodlots, with each woodlot following its own seasonal sequence of occupancy. Some woodlots were occupied in one season only, while others were occupied all year round. Woodlots used for wintering were not necessarily well populated in other seasons (Fig. 5.1) (Kozakiewicz and Gortat 1994).

The spatiotemporal dynamics of a landscape seems to be the crucial factor shaping the space-use pattern of small mammals. This conclusion is supported by a still valuable paper by Southwood (1977), who discussed strategies for reproductive success in habitats that are changeable both in space and in time. Considering the choices open to any organism with respect to its breeding success, Southwood discussed the simplest two-by-two matrix, with "now" and "later" points on the time axis, and "here" and "elsewhere" points

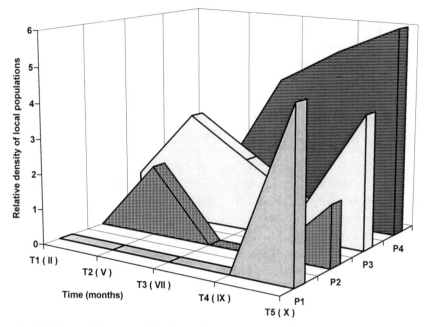

Fig. 5.1. Seasonal dynamics of relative numbers of four selected local bank vole populations (P1-P4) in Polish woodlots. The breeding season in this region is from April (IV) to September (IX). (Redrawn from Kozakiewicz and Gortat 1994, with permission.)

on the space axis. He then distinguished the behavioral-physiological tactics of immediate breeding ("here" and "now"), dormancy ("here" but "later"), dispersal ("elsewhere" and "now"), and the fourth combination ("elsewhere" and "later").

Most authors that study small-mammal movements suggest that individuals that move long distances, and especially those that are nomads, are "losers" or represent a "worse" part of a population (Gliwicz 1989). We reject this notion because we believe that long-distance movements and nomadism are adaptations to landscape variability in space and time. Liro and Szacki (1987) found no differences in mean size, weight, sex ratio, or sexual activity between individuals that were found to be mobile and the rest of the population. At least in heterogeneous landscapes, mobility and nomadism seem to be common traits and may not signal low individual fitness.

Strategies for Survival in Heterogeneous Landscapes

As mentioned earlier, natural systems at any scale of resolution are mosaics of patches. These mosaics are never static: their elements are in constant spatial and temporal flux. Thus, it may be supposed that numerous animal species follow these spatiotemporal habitat variations, continually selecting different patches of habitat. It seems, therefore, that searching for and finding the best-quality patches of habitat that can satisfy an individual's needs may be of great importance to the survival of many species in a mosaic landscape.

In landscape mosaics, the probability that local conditions will worsen at any time is high, so the chances of long-term survival in a particular patch are low. Under such conditions, a single small patch of temporarily suitable habitat cannot satisfy all life requirements of a species and therefore cannot support a stable, viable population. Species must adapt behaviorally to spatial and temporal uncertainties, and at least two strategies for survival can be distinguished: (1) the *sedentary strategy* of being passive and waiting for better days, and (2) the *high spatial activity strategy* of searching for various resources that are dissipated in space and changeable over time. Three tactics of the active strategy variety lead to three space-use patterns (Fig. 5.2): (1) dispersal (finding a new patch, staying there, and breeding early); (2) enlarging as much as possible the range of individual movements in order to cover in routine daily activity more than one habitat patch; and (3) active selection of the best patches of habitat according to present needs, dispersal abilities, and current conditions.

The third type of spatial behavior is opportunistic and serially nomadic, while the first two types are more conservative and sedentary. The strategy of moving freely and choosing the best habitat patches requires high fecundity, a short generation time, and a well-developed ability to travel relatively long distances in a short time. In several species of small mammals, a considerable proportion of individuals in a population can travel across the landscape (Kozakiewicz et al. 1993; Liro and Szacki 1987; Szacki and Liro 1991), which suggests that searching for the best habitat patches, establishing temporary residence, and breeding successfully is a common tactic. Such a serially nomadic existence provides a good opportunity to satisfy all life requirements of the species in a proper time frame.

Which strategy a given population uses depends on the species and the type of landscape it occupies. As mentioned earlier, it seems that the greater the degree of fragmentation, the longer the distance of small-mammal movements. This poses the question of whether or not the trait for great mobility is common among different species of small mammals.

It also seems that among small mammals there are one-habitat species restricted to special vegetation types and multihabitat species (alpha and gamma

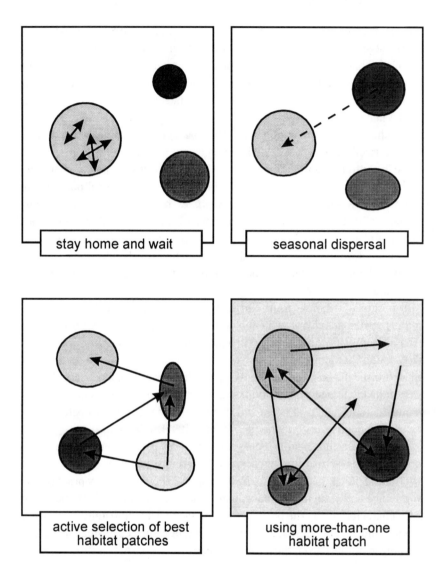

Fig. 5.2. Four patterns of space use by animals living in patchy environments. Arrows indicate possible movements of individuals. Different intensities of shading indicate inhabitable (suitable) patches of various kinds, and white areas indicate uninhabitable (unsuitable) patches.

species, respectively, according to Harris and Kangas 1988). Individuals of specialist species would move from patch to patch, while generalists, being equally mobile, could use an array of different habitats. According to some authors, *Clethrionomys* species are much more attached to one site than many *Apodemus* species, which disperse seasonally in search of resources (Gliwicz 1986; Viitala and Hoffmeyer 1985).

Bauchau and Le Boulengé (1991) describe a good example of two species characterized by different spatial strategies but coexisting in a mosaic landscape in central Belgium. The bank vole is recognized as a species characterized by strict habitat requirements and relatively low dispersal potential. In contrast, the wood mouse, *Apodemus sylvaticus,* has been shown to be a habitat generalist and opportunist exhibiting high dispersal ability. Bank voles were not observed outside woody habitats. They were present in larger and less isolated woodlots, while wood mice occupied nearly all woodlots in the study and moved frequently between habitat patches (Bauchau and Le Boulengé 1991).

In a similar example from North America, Wegner and Henein (1991) studied spatial activity of white-footed mice (*Peromyscus leucopus*) and eastern chipmunks (*Tamias striatus*) in a farmland mosaic near Ottawa, Ontario. They found that individual *Peromyscus* could nest in agricultural fields, feed opportunistically on grain, and move up to 2 km across the landscape. In contrast, individual eastern chipmunks, a woodland specialist species, did not venture into agricultural fields and were strongly restricted to patchily distributed wooded areas (Wegner and Henein 1991).

We conclude that neither bank voles nor eastern chipmunks can travel long distances across the landscape or, therefore, actively select the best patches of habitat. In contrast to the high spatial activity strategy practiced by wood mice and white-footed mice, bank voles and eastern chipmunks seem to be characterized by the patch restriction tactic. Unfortunately, there is not enough information on spatial behavior in enough species of small mammals to safely generalize from these two examples.

Small-Mammal Mobility and the Existing Theories of Populations

In many cases, findings of long-distance movements of small mammals seem to conflict with existing theory. Neither nomadic existence nor the incorporation of different habitats with extensive home ranges fits source-sink models (leaving aside differences concerning the term *sink*). In addition, metapopulation theory is often not applicable to spatial behavior of small mammals in heterogeneous landscapes.

In spite of (or perhaps because of) many papers dedicated to the subject, metapopulation theory is still a rather vague term (Hanski and Gilpin 1991). If we assume that a metapopulation requires living in patches with an adequate degree of isolation (not too low and not too high), then subpopulations of animals that "stay and wait" or migrate occasionally to other patches do form a metapopulation. When animals wander freely from patch to patch or when their great range of movements covers different elements of a landscape, there is no isolation between subpopulations, so there are no extinctions or subsequent recolonizations. Then, the population structure becomes that of a "patchy population" (Lidicker, Chap. 1), and metapopulation theory fails to describe space use adequately.

Extensive spatial activity of small mammals causes difficulties in population theory based on the assumption that small mammals are very much attached to a home site and have stable home ranges that can be shifted only a few meters during one season (e.g., Bujalska 1990). Theories of population regulation based on stable (and sometimes actively defended) home ranges resulting from strong social interactions are not applicable to animals moving, say, a kilometer a day. Social interactions are inevitably weaker when animals are highly mobile, and extrinsic factors become more important. Long-distance movements must be risky, and mortality during them may maintain population numbers below carrying capacity. Such a mechanism can lead to the stability of numbers in heterogeneous environments. Also, predators seem to play a more important role when the landscape heterogeneity increases (Andrén and Angelstam 1988; Andrén et al. 1985; Goszczyński 1985; Hansson 1984; Lidicker, Chap. 1; Oksanen and Schneider, Chap. 7).

Neither the concept of small home ranges where animals meet all their needs nor methods of density estimation used so far can be applied in situations where nomadism is rampant. As is sometimes pointed out, our anthropocentric way of perceiving the world may severely bias our interpretation of the spatial orientation of small mammals. For example, we know that rodents rely largely on olfaction rather than on vision. Because the air in a field moves along narrow and winding channels (Francevic 1986), rodents may perceive the landscape as a network of olfactory paths and not the area as a whole. Social relations based on olfactory signals would be rather indirect.

Thus, the assumption that long-distance movements are common among small mammals, at least in heterogeneous environments (and perhaps for rodents there are no homogeneous environments), requires major changes in the present theory of animal populations. In fact, this theory should be recreated from the very beginning, even though the task may require a great deal of investigation and appears to be very difficult.

Consequences for Conservation

If animals are to move long distances, two conditions must be met: they must be motivated, and they must be capable of making such journeys. We have tried to show that in heterogeneous landscapes they are motivated. Human activity, however, such as building roads, power lines, or railways, creates barriers that prevent small mammals from covering long distances no matter how much they are motivated. Such barriers must have serious implications for population organization (Kozakiewicz 1983). Szacki (1987) studied small-mammal populations in patches of habitat and observed that populations of *C. glareolus,* individuals of which were not able to traverse a strip of meadow between patches, were less stable than a population of *A. flavicollis,* which moved freely across the landscape.

Habitat barriers can be classified into three categories with respect to the scale of their effects: those that affect movements of single individuals (individual home range level) but do not cause discontinuities within the population; those that separate local populations but do not preclude all movements (metapopulation level); and those that are effective enough to decrease gene flow between local populations to the level that allows interdemic genetic differentiation (evolutionary deme) (Dobrowolski et al. 1993). Moreover, the interpatch connectivity can be modified by a barrier "filtering" effect. According to Kozakiewicz (1993), habitat barriers may act as "filters" that stop some individuals and allow others to pass through, depending on their ability to move. Such filters can play an important role in structuring small populations isolated by habitat barriers (Kozakiewicz and Jurasińska 1989). By differentially filtering out species, habitat barriers can also have an important influence on species distribution across fragmented landscapes (Kozakiewicz 1993). This reasoning suggests that in mosaic landscapes, the functional connectivity (Merriam 1984) is a crucial parameter, since free movements of organisms between habitat patches can influence the demographic properties and stability of single populations as well as the dynamics of the whole mosaic system.

Small mammals (and there is no evidence that this does not apply to other mammals as well) seem to use strict routes when moving across a landscape. Several authors have stressed the role of corridors for animal movements (Fahrig et al. 1983; Hansson 1987; Henein and Merriam 1990; Merriam 1991; Merriam and Lanoue 1990). The role of corridors in conservation of endangered species as well as in maintenance of biodiversity is also widely recognized and discussed (Hansson 1987; Harris 1985, 1988; Harris and Gallagher 1989; Simberloff and Cox 1987; and many others). Liro and Szacki (1987) and Szacki and Liro (1991) showed that despite the great mobility of small mammals, the distribution of catches along traplines is very uneven, even when traps are only a few meters apart and the landscape seems relatively homogeneous. This suggests not only that different habitats are used as movement corridors in different ways, but also that some trails are used as movement routes. Unfortunately, these corridors and trails are often not easily perceived by humans. There may be functional corridors, even though difficult for us to recognize, as well as structural ones that are easily recognized strips of vegetation that can be easily distinguished from the surroundings. Considerable problems may arise in habitat conservation efforts where functional corridors are not discovered in a visual survey. More detailed studies covering the requirements of particular species and not of wildlife as a whole often may be needed.

Regarding the temporal dynamics of species requirements and spatiotemporal changeability of environments, we suggest that mosaics of a variety of habitats may benefit generalist and highly mobile species. Individuals of such multihabitat species probably can move freely among patches of habitat and select them according to temporal changes in their quality. Habitat specialists, in contrast, may go extinct if dispersal is poor enough that preferred habitats are not found. Thus, the more heterogeneous the habitat mosaic, the higher the expected proportion of generalist species and the lower the proportion of habitat specialists for which preferred habitats may be too scattered.

Another problem for conservation is the existence of so-called key habitats (Kozakiewicz 1993, 1995). In some seasons or in particular stages of a life cycle, species' needs can be very special, and filling these needs can be critical for population survival. Kozakiewicz (1993, 1995) considers habitats in which such needs can be met "key habitats" for population persistence. In conservation biology they are often called "critical habitats" (Harris and Kangas 1988). Such key habitats can be spatially scattered across landscapes so as to benefit some highly mobile species. Many less mobile species, however, may go extinct in such landscapes. For conservation purposes, therefore, it may be necessary in some cases to investigate whether individual patches of a landscape are critical for some species of interest, even if for only a short time.

Conclusions

Recently, evidence has been mounting indicating considerable mobility of small mammals of various species. The greater the heterogeneity of the landscape, the greater the apparent mobility of animals.

Animals living in landscapes that are heterogeneous in space and changeable over time may display four types of spatial behavior, depending on species requirements: (1) the sedentary strategy; (2) incorporation of more than one patch into a home range; (3) seasonal dispersal and use of more than one habitat patch; (4) active selection (serial nomadism) of the best patches (Fig. 5.2). Active selection seems to be the most common strategy for survival in heterogeneous landscapes. Such high mobility of animals calls into question the existing understanding of population regulation based on small and stable home ranges and strong bonds among individuals.

Finally, as regards conservation, the most important problems related to high spatial activity of animals are ensuring spatial connectivity and identifying and preserving key habitats that hold special significance for the survival of many species. Creating and maintaining functional corridors that are not necessarily structures recognized by humans is especially important for specialist species otherwise unable to travel long distances across a landscape. And the conservation of particular species may require the conservation of particular key habitats.

Summary

The inclination and ability of organisms to move among patches are critical to their success in heterogeneous landscapes. Yet few data document how far individual small mammals can move on a regular basis, how they respond to patch edges, how they are influenced by the interpatch matrix, how they navigate over large areas, or even whether patches of different kinds are necessary or preferred in various seasons. We present several examples of both North American and European species living in various mosaic landscapes. The data presented suggest that the basic home range concept that is so fundamental to much of population dynamics and dispersal theory may have to be modified for strongly subdivided habitats. Fragmentation may also modify community composition so as to favor vagile and generalist species. A review of the small-mammal literature combined with our evidence suggests life history strategies that are conducive to survival in fragmented habitats. The relevance of these insights to conservation biology is discussed.

Literature Cited

Andrén, H., and P. Angelstam. 1988. Elevated predation rates as an edge effect in habitat islands: Experimental evidence. *Ecology* 69:544–7.

Andrén, H., P. Angelstam, E. Lindström, and P. Widén. 1985. Differences in predation pressure in relation to habitat fragmentation: An experiment. *Oikos* 45:273–7.

Andrzejewski, R. 1963. Processes of incoming, settlement and disappearance of individuals and variations in the numbers of small rodents. *Acta Theriol.* 7:169–213.

Andrzejewski, R., and J. Babińska-Werka. 1986. Bank vole populations: Are their densities really high and individual home range small? *Acta Theriol.* 31:409–22.

Bauchau, V., and E. Le Boulengé. 1991. Population biology of woodland rodents in a patchy landscape. Pages 275–83 *in* Le Berre, M., and L. Le Guelte (eds.), *Le Rongeur et l'Espace*. Paris: R. Chabaud. 362 pp.

Begon, M., J. L. Harper, and C. R. Townsend. 1990. *Ecology; Individuals, Populations and Communities*, 2d ed. Oxford: Blackwell Scientific Publications. 945 pp.

Bock, E., and J. Babińska. 1967. The influence of prebaiting on the catches of rodents. *Small Mammal Newsl.* 3:18–9.

Brown, J. S., and N. B. Pavlovic. 1992. Evolution in heterogeneous environments: Effects of migration on habitat specialization. *Evol. Ecol.* 6:360–82.

Buchalczyk, T., and Z. Pucek. 1968. Estimation of the numbers of *Microtus oeconomus* using the standard minimum method. *Acta Theriol.* 13:461–82.

Bujalska, G. 1990. Social system of the bank vole, *Clethrionomys glareolus*. Pages 155–68 *in* Tamarin, R. H., R. S. Ostfeld, S. R. Pugh, and G. Bujalska (eds.), *Social Systems and Population Cycles in Voles*. Basel, Switzerland: Birkhäuser Verlag. 229 pp.

Clark, B. K., G. A. Kaufman, E. J. Finck, and S. S. Hand. 1988. Long-distance movements by *Reithrodontomys megalotis* in tallgrass prairie. *Am. Midl. Nat.* 120:276–81.

Dickman, C. R., and C. P. Doncaster. 1989. The ecology of small mammals in urban habitats. II. Demography and dispersal. *J. Anim. Ecol.* 58:119–27.

Dobrowolski, K., A. Banach, A. Kozakiewicz, and M. Kozakiewicz. 1993. Effect of habitat barriers on animal populations and communities in heterogeneous landscapes. Pages 61–70 *in* Bunce, R. G. H., L. Ryszkowski, and M. G. Paoletti (eds.), *Landscape Ecology and Agroecosystems*. Boca Raton, Fla.: Lewis Publishers. 241 pp.

Dunning, J. B., B. J. Danielson, and H. R. Pulliam. 1992. Ecological processes that affect populations in complex landscapes. *Oikos* 65:169–75.

Fahrig, L. 1990. Interacting effects of disturbance and dispersal on individual selection and population stability. *Comments Theor. Biol.* 1:275–97.

Fahrig, L., L. P. Lefkovitch, and H. G. Merriam. 1983. Population stability in a patchy environment. Pages 61–7 *in* Lauenroth, W. K., G. V. Skogerboe, and M. Flug (eds.), *Analysis of Ecological Systems: State-of-the-Art in Ecological Modelling*. New York: Elsevier. 992 pp.

Francevic, L. I. 1986. *Prostranstvennaya Orientacia Zivotnych*. Kiev: Naukowa Dumka. 196 pp. (In Russian)

Gliwicz, J. 1986. Migracja w populacjach gryzoni. *Wiad. Ekol.* 32:137–54. (English summary)

Gliwicz, J. 1989. Individuals and populations of the bank vole in optimal, suboptimal and insular habitats. *J. Anim. Ecol.* 58:237–47.

Gliwicz, J. 1992. Patterns of dispersal in non-cyclic populations of small rodents. Pages 147–59 *in* Stenseth, N. C., and W. Z. Lidicker Jr. (eds.), *Animal Dispersal: Small Mammals as a Model*. London: Chapman & Hall. 365 pp.

Goszczyński, J. 1985. The effect of structural differentiation of ecological landscape on the predator-prey interaction. *Publ. Warsaw Agric. Univ. Treatises Monogr.* 46:1–80. (In Polish with English summary)

Hanski, I., and M. Gilpin. 1991. Metapopulation dynamics: Brief history and conceptual domain. *Biol. J. Linn. Soc.* 42:3–16.

Hansson, L. 1984. Predation as a factor causing extended low densities in microtine rodents. *Oikos* 43:255–6.

Hansson, L. 1987. Dispersal routes of small mammals at an abandoned field in central Sweden. *Holarctic Ecol.* 10:154–9.

Harris, L. D. 1985. *Conservation Corridors: A Highway System for Wildlife.* ENFO Publ. 85–5. Winter Park, Fla.: Environmental Information Center. 10 pp.

Harris, L. D. 1988. Landscape linkages: The dispersal corridor approach to wildlife conservation. *Trans. North Am. Wildl. Nat. Resour. Conf.* 53:595–607.

Harris, L. D., and P. B. Gallagher. 1989. New initiatives for wildlife conservation. The need for movement corridors. Pages 11–34 *in* Mackintosh, G. (ed.), *In Defense of Wildlife: Preserving Communities and Corridors.* Washington, D.C.: Defenders of Wildlife. 96 pp.

Harris, L. D., and P. Kangas. 1988. Reconsideration of the habitat concept. *Trans. North Am. Wildl. Nat. Resour. Conf.* 53:137–44.

Harrison, S. 1991. Local extinction in a metapopulation context: An empirical evaluation. *Biol. J. Linn. Soc.* 42:73–88.

Henein, K., and G. Merriam. 1990. The elements of connectivity where corridor quality is variable. *Landscape Ecol.* 4:157–70.

Kolasa, J., and C. D. Rollo. 1991. Introduction: The heterogeneity of heterogeneity. Pages 1–23 *in* Kolasa, J., and S. T. A. Pickett (eds.), *Ecological Heterogeneity.* Ecological Studies 86. New York: Springer-Verlag. 332 pp.

Kozakiewicz, A., R. Banaszewska, and M. Kozakiewicz. 1994. Track registry—A method for studying spatial activity of small mammals. *Pol. Ecol. Stud.* 20:199–204.

Kozakiewicz, M. 1983. Environmental and ecological effects of artificial division of the population area. Pages 22–3 *in* Calhoun, J. B. (ed.), *Environment and Population: Problems of Adaptation.* New York: Praeger Publishers. 486 pp.

Kozakiewicz, M. 1993. Habitat isolation and ecological barriers—The effect on small mammal populations and communities. *Acta Theriol.* 38:1–30.

Kozakiewicz, M. 1995. Resource tracking in space and time. Pages 136–48 *in* Hansson, L., L. Fahrig, and G. Merriam (eds.), *Mosaic Landscapes and Ecological Processes.* London: Chapman & Hall. 356 pp.

Kozakiewicz, M., and T. Gortat. 1994. Abundance and seasonal dynamics of bank voles in a patchy agricultural landscape. *Pol. Ecol. Stud.* 20:209-14.

Kozakiewicz, M., and E. Jurasińska. 1989. The role of habitat barriers in woodlot recolonization by small mammals. *Holarctic Ecol.* 12:106–11.

Kozakiewicz, M., A. Kozakiewicz, A. Łukowski, and T. Gortat. 1993. Use of space by bank voles (*Clethrionomys glareolus*) in a Polish farm landscape. *Landscape Ecol.* 8:19–24.

Krohne, D. T., and A. B. Burgin. 1990. The scale of demographic heterogeneity in a population of *Peromyscus leucopus. Oecologia* 82:97–101.

Lidicker, W. Z., Jr. 1975. The role of dispersal in the demography of small mammals. Pages 103–28 *in* Golley, F. B., K. Petrusewicz, and L. Ryszkowski (eds.), *Small Mammals: Their Productivity and Population Dynamics.* Cambridge: Cambridge University Press. 451 pp.

Lidicker, W. Z., Jr. 1988. Solving the enigma of microtine "cycles." *J. Mammal.* 69:225–35.

Lidicker, W. Z., Jr., and J. L. Patton. 1987. Patterns of dispersal and genetic structure in populations of small rodents. Pages 144–61 *in* Chepko-Sade, B. D., and Z. Tang Halpin (eds.), *Mammalian Dispersal Patterns; The Effects of Social Structure on Population Genetics.* Chicago: University of Chicago Press. 342 pp.

Lidicker, W. Z., Jr., and N. C. Stenseth. 1992. To disperse or not to disperse: Who does it and why? Pages 21–36 *in* Stenseth, N. C., and W. Z. Lidicker Jr. (eds.), *Animal Dispersal; Small Mammals as a Model.* London: Chapman & Hall. 365 pp.

Liro, A., and J. Szacki. 1987. Movements of field mice *Apodemus agrarius* (Pallas) in a suburban mosaic of habitats. *Oecologia* 74:438–40.

Lord, J. M., and D. A. Norton. 1990. Scale and the spatial context of fragmentation. *Conserv. Biol.* 4:197–202.

Merriam, G. 1984. Connectivity: A fundamental ecological characteristic of landscape pattern. Pages 5–15 *in Proceedings of the First International Seminar on Methodology in Landscape Ecological Research and Planning.* Roskilde, Denmark: International Association for Landscape Ecology. 118 pp.

Merriam, G. 1988. Landscape dynamics in farmland. *Trends Ecol. Evol.* 3:16–20.

Merriam, G. 1990. Ecological processes in the time and space of farmland mosaics. Pages 121–33 *in* Zonnefeld, I. S., and R. T. T. Forman (eds.), *Changing Landscapes: An Ecological Perspective.* New York: Springer-Verlag. 286 pp.

Merriam, G. 1991. Corridors and connectivity: Animal populations in heterogeneous environments. Pages 133–42 *in* Saunders, D. A., and R. J. Hobbs (eds.), *Nature Conservation 2: The Role of Corridors.* Chipping Norton, N.S.W., Australia: Surrey Beatty & Sons. 442 pp.

Merriam, G., and A. Lanoue. 1990. Corridor use by small mammals: Field measurement for three experimental types of *Peromyscus leucopus. Landscape Ecol.* 4:123–31.

Milne, B. T. 1991. Heterogeneity as a multiscale characteristic of landscapes. Pages 69–84 *in* Kolasa, J., and S. T. A. Pickett (eds.), *Ecological Heterogeneity.* Ecological Studies 86. New York: Springer-Verlag. 332 pp.

Morris, D. W. 1992. Scales and costs of habitat selection in heterogeneous landscapes. *Evol. Ecol.* 6:412–32.

Sheppe, W. 1967. The effect of live-trapping on the movements of *Peromyscus. Am. Midl. Nat.* 78:471–81.

Simberloff, D., and J. Cox. 1987. Consequences and costs of conservation corridors. *Conserv. Biol.* 1:63–71.

Southwood, T. R. E. 1977. Habitat, the templet for ecological strategies? *J. Anim. Ecol.* 46:337–65.

Stenseth, N. C., and W. Z. Lidicker Jr. 1992. The study of dispersal: A conceptual guide. Pages 5–20 *in* Stenseth, N. C., and W. Z. Lidicker Jr. (eds.), *Animal Dispersal; Small Mammals as a Model.* London: Chapman & Hall. 365 pp.

Szacki, J. 1987. Ecological corridor as a factor determining the structure and organization of bank vole population. *Acta Theriol.* 32:31–44.

Szacki, J., J. Babińska-Werka, and A. Liro. 1993. The influence of landscape spatial structure on small mammal movements. *Acta Theriol.* 38:113–24.

Szacki, J., and A. Liro. 1991. Movements of small mammals in the heterogeneous landscape. *Landscape Ecol.* 5:219–24.

Tew, T. 1988. The ecology of the European wood mouse (*Apodemus sylvaticus*) on British farmland. Page 103 *in Proceedings of the Second International Behavioural Ecology Conference, Vancouver.*

Viitala, J., and I. Hoffmeyer. 1985. Social organization in *Clethrionomys* compared with *Microtus* and *Apodemus:* Social odours and biological effects. *Ann. Zool. Fenn.* 22:359–71.

von Somsook, S., and H. M. Steiner. 1991. Zur Grosse des Aktionraumes von *Microtus arvalis* (Pallas, 1779). *Z. Saugetierkunde* 56:200–6.

Wegner, J., and K. Henein. 1991. Strategies for survival: White-footed mice and eastern chipmunks in an agricultural landscape. Page 90 *in Proceedings of the World Congress of*

Landscape Ecology, Ottawa, Canada. International Association for Landscape Ecology. 103 pp.

Wegner, J., and G. Merriam. 1990. Use of spatial elements in a farmland mosaic by a woodland rodent. *Biol. Conserv.* 54:263–76.

Wiens, J. A. 1976. Population responses to patchy environments. *Annu. Rev. Ecol. Syst.* 17:81–120.

Wiens, J. A. 1985. Vertebrate responses to environmental patchiness in arid and semiarid ecosystems. Pages 169–93 *in* Pickett, S. T. A., and P. S. White (eds.), *The Ecology of Natural Disturbance and Patch Dynamics.* San Diego: Academic Press. 472 pp.

Wiens, J. A., J. F. Addicott, T. J. Case, and J. Diamond. 1986. Overview: The importance of spatial and temporal scale in ecological investigation. Pages 145–53 *in* Diamond, J., and T. J. Case (eds.), *Community Ecology.* New York: Harper & Row. 665 pp.

Wolton, R. J. 1985. The ranging and nesting behaviour of wood mice, *Apodemus sylvaticus* (Rodentia, Muridae). *J. Zool. Ser. A* 206:203–24.

6

Martes Populations as Indicators of Habitat Spatial Patterns: The Need for a Multiscale Approach

John A. Bissonette and Sim Broekhuizen

Martens are medium-sized carnivores that range across areas of up to several thousand hectares and select core use areas from the larger landscape. Because of this characteristic, they are good indicators of habitat quality. Both American and European species appear to avoid open areas. As a consequence, the most obvious characteristics that appear to influence marten numbers and distribution are habitat fragmentation and prey availability.

To distinguish the effects of larger-scale landscape influences from those of agents operating on smaller spatial and temporal scales, marten populations must be studied at multiple scales. Otherwise, data obtained from few animals over short periods and on a small spatial scale may not be generalizable to larger scales; that is, qualitatively different results will be obtained at larger scales if data are transmuted when scaled upward.

In this chapter, we synthesize our experiences with European and North American martens in order to test our intuition that scale effects are extremely important. We feel that it is imperative for researchers to give explicit a priori consideration to both temporal and spatial scales before conservation studies are conducted.

What Is Good Habitat?

American Marten

Martens in North America (*Martes americana*) are found primarily in older-growth coniferous forests, particularly spruce and fir. In Newfoundland, the densest populations were found in balsam fir–white birch (*Abies balsamea–Betula papyrifera*) stands with more than 76% overstory (Bateman 1986). Bissonette et al. (1988) found that overhead cover, although a correlate of older-growth forests in Newfoundland, did not appear to be absolutely required. G. S. Drew and J. A. Bissonette (unpub. data) observed martens foraging through high-density defoliated stands with abundant dead and fallen woody debris on the forest floor. Koehler (1975) found that habitat use during the winter in Idaho varied, depending on the severity of the winter: during severe winters, marten activity was greatest in mesic older-growth forests more than 100 years old, with overstory canopy cover of at least 30%; in milder winters, habitat use was less restricted.

Although martens appear to prefer dense, undisturbed older-growth coniferous forest, Buskirk (1983) reported that there were many such areas in Alaska where martens were absent. *Martes* populations are also found in mixed forests. Hargis and McCullough (1984) and Martin (1987) in California and Sherburne and Bissonette (1993) in Wyoming found that martens made significant use of lodgepole pine (*Pinus contorta*) forests. Hargis and McCullough (1984) also found that martens preferred areas with overhead cover less than 3 m above the snow both when traveling and when resting. Raine (1983) reported that soft snow did not appear to hinder marten movements as much as it did movements of fishers (*M. pennanti*). G. S. Drew (pers. comm.) found that American martens most often approached an edge or clearing from within the interior forest rather than by following an edge. However, Spencer et al. (1983) reported that martens in the northern Sierra Nevada of California were found most often within 30–400 m of a meadow. Major (1979) found high use during summer of regenerating clear-cuts, where martens foraged on raspberries (*Rubus idaeus*). Further, marten tracks were found in several shrub habitats, although in much reduced numbers. Marten sign in these marginal (nonforest) habitats is likely the result of juvenile movements. Several authors (Bissonette et al. 1988; Hargis 1981) have reported that martens will not cross wide open areas. Bissonette et al. (1988) found that young of the year were much more likely than adults to use thinned or clear-cut openings. There appear to be seasonal elevational shifts in marten habitat use. Buskirk (1983) reported that martens moved to higher elevations in spring and lower elevations in fall.

American martens appear to prefer older-growth forests for at least three reasons: protection from predation, homeothermic management, and access to a prey base. Martens appear more closely confined to older-growth forests during winter than during summer, when vegetation growth provides more vertical and horizontal cover. During winter, the coarse woody debris (CWD) associated with older forests intercepts snowfall, creating holes, cavities, and interstitial spaces that martens use as access for homeothermic and prey purposes. Homeothermic management and subnivean access are winter-related phenomena.

Sherburne (1992) reported that martens in Yellowstone National Park used lodgepole pine habitats even after they had been burned, but their manner of use indicated that predation may be a potent factor in marten habitat selection. Martens used canopy-burned stands during the winter, and their tracks indicated that they were hunting through the area. These habitats still had unburned CWD on the forest floor that provided subnivean access to the prey base. However, marten travel through surface-burned stands was characterized by relatively straight-line movements between standing trees, with little or no evidence of hunting. The martens appeared to be using the trees for security. In effect, trees provided a vertical escape route from terrestrial predators, including both coyotes (*Canis latrans*) and red fox (*Vulpes vulpes*), which have been documented to kill martens in Yellowstone (S. S. Sherburne, pers. comm.). G. S. Drew and J. A. Bissonette (unpub. data) have shown that martens use insect-defoliated stands in Newfoundland in a similar manner. Further, in field experiments, they placed baited boxes in the forest and 25, 50, and 75 m from the forest edge extending into an open meadow or clear-cut area. Martens visited the bait box 25 m from the forest edge but used the 50 m box only if they could find an intervening tree that effectively shortened the distance. They never visited the 75 m bait box. These results indicate clearly that marten habitat choice is related to predation pressures.

European Pine Marten

European pine martens (*M. martes*) show a variable habitat preference. Like American martens, pine martens most often are associated with forested areas. However, their preferred forest type differs with latitude and altitude (Heptner et al. 1974). Preference for older mixed-conifer cover is mentioned for Finland (Pulliainen 1981a), Scotland (Balharry 1984), Denmark (Degnn and Jensen 1977), and the Czech Republic (Nesvadbova and Zejda 1984); preference for mixed deciduous woodland and shrubland is reported from Ireland (O'Sullivan 1983), the Swiss Jura Mountains (Marchesi 1989), and Czechoslovakia (Pelikán and Vačkař 1978); and preference for pine and shrubland habitats is reported for Minorca, Spain (Clevenger 1993).

Although pine martens are reported to avoid open areas (Storch et al. 1990), some authors also mention their occurrence in open habitats (Clevenger 1993) and cultivated areas with only small woodlots (Dolch 1991; Marchesi 1989; G. J. D. M. Müskens and S. Broekhuizen, unpub. data). Occasionally, they are seen in the vicinity of houses (Marchesi 1989), and sometimes maternal dens are found under roofs of sheds and summerhouses (G. J. D. M. Müskens and S. Broekhuizen, unpub. data).

Stone Marten

The stone or beech marten (*M. foina*) is a more anthropophilic species inhabiting rural and urban habitats (Herrmann 1987; Holišova and Obrtel 1981; Müskens et al. 1989; Nicht 1969; Rasmussen et al. 1986; Skirnisson 1986; Tester 1986), but in some parts of its area of distribution it forages mainly in forested environments while avoiding human surroundings (Delibes 1978; Lachat Feller 1993).

Home Range Use

Martens' use of their home range is influenced directly and indirectly by snow cover and mild weather (Koehler 1975; Koehler and Hornocker 1977), prey base abundance, and disturbance by fire and habitat alteration. Koehler and Hornocker (1977) reported that during winters with heavy snow, marten activity was concentrated in conifers with high stand density, whereas during milder winters, stand density appeared to have little influence—martens were able to forage throughout their home range. The availability and abundance of seasonal food clearly influence use of space. Clark and Campbell (1979), Major (1979), Soutiere (1979), and Steventon and Major (1982) reported that martens in late summer and autumn used more open parts of their home range to find raspberries (*Rubus* spp.). Thompson and Colgan (1990) and Bissonette et al. (1988, 1989) reported increased movement of martens due to food shortages.

Martell and Radvanyi (1977) and Martell (1984) pointed out that habitat alteration influences the species composition of the prey base. Clark and Campbell (1979) showed differences in the prey base associated with logging operations, principally on xeric sites where deer mice (*Peromyscus*) were more abundant. Both Clough (1987) and Stephenson (1984) suggested that *Peromyscus* tend to dominate the small-mammal community after a fire or other disturbance. Monthey and Soutiere (1985) showed that the total small-mammal community increased following harvesting in softwood stands. However, Soutiere (1979) claimed that in Maine martens seldom eat deer mice.

Marten home ranges are rarely homogeneous in structure or productivity. All investigators report use of core areas. Core areas may change through the seasons and from year to year, due to fluctuating prey density and the defoliation of herbs and shrubs. Marchesi (1989) tracked a female *M. martes* for 10 months and found that activity was more widespread during winter and summer than in spring and autumn, when the animal concentrated particularly on fruit.

Resting Sites

Many authors have reported the importance to American martens of forest floor structure for resting sites during the summer. Simon (1980) found that martens oriented toward special habitat components, including snags, dead and fallen woody debris, meadow-forest edges, and streamsides. Martin and Barrett (1983) found that 23% of all dens in the Sierra Nevada in California were associated with snags. They also reported that 74% of 155 marten resting sites were found in stumps, snags, logs, or tree canopy (Martin and Barrett 1991). However, Wynne (1981) reported that in Maine all summer resting sites used by males were in tree canopies. Clearly, resting site selection is not nearly as constrained in summer as during winter. Steventon and Major (1982) documented repeated use of certain resting sites in Maine, although each site was used by only one animal. Twenty-four of 155 sites (16%) in California were reused a total of 48 times (Martin and Barrett 1991).

During winter, *M. americana* appear to require CWD primarily for prey access and for thermoregulation (Sherburne and Bissonette 1993, 1994). Resting sites in this season are primarily subnivean and associated with CWD (Buskirk et al. 1987). Buskirk et al. (1989) speculated that use of winter resting sites might be associated with the need for thermal cover, which may explain why martens used older-growth forests in winter in the Rocky Mountains. Buskirk et al. (1988) showed that martens selected the warmest sites during the coldest weather, and they concluded that site selection was based primarily on the thermal properties of the sites. They and Worthen and Kilgore (1981) found that martens have relatively high lower critical temperatures and suggested that the animals coordinated body temperature, use of thermal cover, and duration of resting episodes as a means of energy management in response to environmental conditions (Buskirk et al. 1988).

Nineteen of 31 subnivean resting sites analyzed by Steventon and Major (1982) were associated with decaying large stumps with natural cavities. Buskirk (1984) and Buskirk et al. (1989) found that most resting sites were in older-growth forests and were associated with red squirrel (*Tamiasciurus hudsonicus*) middens. Martens seemed to prefer cavities in decayed wood below the snow with squirrel cone caches during winter. Spencer (1987) reported that martens used live trees as resting sites when snow cover was less than complete, but only subnivean resting sites were used during periods of complete snow cover. Snags and stumps were used more than were downed logs. Spencer also identified important factors in rest site selection. During periods of incomplete snow cover, martens used live trees (mostly lodgepole pine) most frequently, although they were used less than expected based on tree availability. Snags were used most relative to availability and were located almost exclusively in large firs that retained most of their bark. During periods of complete snow cover, all resting sites were subnivean, mostly in snags and decayed stumps rather than under logs. Subnivean sites were reused, sometimes for several days. Most juveniles rested in trees other than spruce, while adult resting sites were evenly divided between spruce or fir and other types. Fidelity to sites was greatest among adults and to sites associated with CWD. Bergerud (1969) suggested that the lack of large denning trees may have precluded habitation by martens of certain forested areas in Newfoundland.

R. W. Threader (unpub. data) used artificial resting sites in controlled laboratory experiments with American marten. On cold days, martens chose insulated sites when given a choice. They could not balance heat loss with metabolic heat production in uninsulated sites.

M. martes choose resting places similar to those chosen by *M. americana.* Marchesi (1989) found that in summer most resting places were located on bird nests and in witches'-broom. In winter, underground cavities, especially those with snow cover, were preferred. During rainy weather, snags and holes of black woodpeckers (*Dryocopus martius*) were preferred.

Van Bostelen and Verhoog (1992) found that in summer, *M. foina* preferred open, uninsulated sites. Stone martens were often found resting on the ground, sometimes without dense vegetation. In winter, insulated sites were chosen. In rural areas, straw piles were favored (Herrmann 1987). Van Bostelen and Verhoog (1992) and Lachat Feller (1993) both noted remarkable differences among individuals in resting site selection under the same circumstances. Simultaneous use of day hides by male and female stone martens has been observed in all seasons (Lammertsma 1992; Müskens et al. 1989).

Population Density and Habitat

The cutting and burning of habitat affects American martens both directly and indirectly and over both the short run and the long run. However, these effects depend on the scale of the changes. Very small openings do not appear to have a negative impact on martens; in fact, they may have a positive effect if they are associated with an increase in the prey base. Larger clear-cuts, however, are clearly detrimental. Bissonette et al. (1988, 1989) found that martens showed strong avoidance of fresh clear-cuts. During logging operations, martens shifted their areas of concentration to other parts of their home range, away from the logging. After the cutting, martens continued to shift their activity to the remaining uncut portions of their home ranges, or they expanded their home ranges into previously unused areas.

Population densities are much reduced in logged forests. Bissonette et al. (1988) showed that logging activities in Newfoundland severely reduced marten populations. Soutiere (1978) documented different population responses to two wood-harvesting methods in Maine. Marten densities in partially harvested forest with basal area reduced by 40% were similar to those in the uncut control (1.2/km²) over the short term. Densities were reduced by two-thirds to 0.4/km² in the clear-cut areas. Steventon and Major (1982) demonstrated that martens' avoidance of clear-cut areas was especially pronounced during winter.

Influences on marten population densities and habitat use also depend on spatial scales. Differences in prey abundance can be localized, resulting in seasonal variation in home range use, or more widespread, causing major disruption in marten spatial dynamics. For example, Bissonette et al. (1988) reported a regionwide prey crash, primarily of *Microtus,* in Newfoundland that resulted in increased mortality of martens and abandonment of their home ranges. Weekly movements increased dramatically.

Although both American and European marten species are able to adapt to different habitats, *M. americana* appears more sensitive to habitat changes and perhaps also to prey abundance. At present, *M. foina* seems less sensitive, especially where it lives in rural or urban habitats. It is remarkable that in many places in western and central Europe, stone marten populations have increased since the late 1950s, in the presence of increasing urbanization (Fig. 6.1). Better conditions—for example, more centrally heated buildings offering better day hides in winter—may be involved, and stone martens may have become more adaptable to a changing rural and urban habitat.

Diet

American martens have been characterized alternatively as specialists and opportunists (Strickland et al. 1982). They prey primarily on small mammals; microtines make up a large part of their diet across North America (Bissonette et al. 1988; Buskirk 1983; Lensink et al. 1955; Murie 1961; Slough et al. 1989; Tucker 1988; Weckwerth and Hawley 1962). However, several authors report that they appear to be opportunistic (Thompson 1986), taking items as

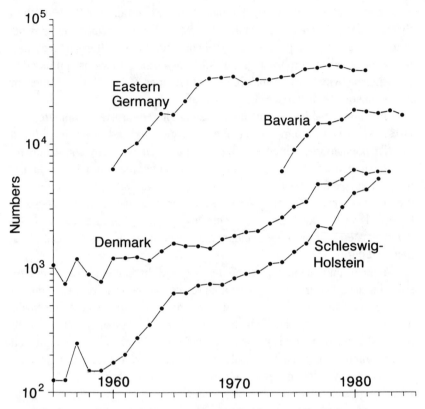

Fig. 6.1. Yearly harvest of stone martens *(Martes foina)* in several European regions, 1955–84.

encountered. The results of reported studies of food habits vary greatly. For example, Newby (1951) reported that arthropods ranked first in the diet, with squirrels accounting for a major portion at certain times. Buskirk (1983) reported strong seasonal components to diet selection. In Newfoundland, the small-mammal prey base is depauperate, with only six species (Bissonette et al. 1988). *Clethrionomys* is absent, and *Microtus pennsylvanicus* is the only microtine present. Nagorsen et al. (1989) reported that more than half of the small mammals in the diet of martens on Vancouver Island, British Columbia, were deer mice *(Peromyscus)*. Buskirk (1983) summarized the general diet preferences of *Martes* across North America as follows: *Microtus, Synaptomys,* and *Phenacomys* are highly preferred; *Clethrionomys, Zapus,* and *Napeozapus* are of intermediate importance; and *Peromyscus* and *Sorex* are of low importance.

Marten population density fluctuates with the density of main prey items for both *M. americana* (Thompson 1987; Weckwerth and Hawley 1962) and *M. foina* (Lachat Feller 1993). Sherburne and Bissonette (1993, 1994) demonstrated that martens used subnivean access sites with higher small-mammal densities significantly more than they used sites with lower densities. Further, sites with a squirrel midden were more heavily used. Clark and Campbell (1979) found no evidence to indicate that food was obtained above ground during winter, except for one occasion when a ruffed grouse (*Bonasa umbellus*) kill site was found.

European martens are opportunistic generalist feeders, consuming a great diversity of animal and vegetable species (*M. martes:* Ansorge 1989b; Marchesi and Mermod 1989; Rzebik-Kowalska 1972; Storch et al. 1990; Warner and O'Sullivan 1982; *M. foina:* Amores 1980; Ansorge 1989a; Delibes 1978; Kalpers 1983; Lachat Feller 1993; Rasmussen and Madsen 1985; Skirnisson 1986). *M. martes* appear to be slightly more carnivorous than *M. foina* (Ansorge 1989b; Grupe and Kruger 1990). In rural areas (Ansorge 1989a; Skirnisson 1986) as well as in urban habitats (Rasmussen and Madsen 1985; Tester 1986), scats and stomach contents from beech martens (*M. foina*) were found to consist of passerines and anthropogenic wastes.

Home Range Dynamics

Home Range Size

Mustelid home range sizes vary with the patchiness of the environment, body mass, individual trophic status (Harestad and Bunnell 1979), and perhaps most important, the productivity of the habitat, that is, prey abundance and availability. Male ranges typically are largest in American martens, although home range size varies considerably among study sites; female ranges appear to vary little (Buskirk and McDonald 1989). Home range size is correlated with body mass for females, but not for males. Males typically weigh about 1.5 times as much as females, and their home ranges are about 1.9 times as large. Measured female home ranges vary from 59 to 2,056 ha, while those of males range from 70 to 2,750 ha (Buskirk and McDonald 1989). Other investigators have reported values for home range size that typically fall within these limits. Interpretation of home range data is hampered by small sample sizes, short data collection periods, lack of information regarding home range history, and lack of associated prey availability data.

Many authors have pointed to the influence of food availability on home range size, but quantitative data are scarce. Marchesi (1989) described the foraging technique of the pine marten as a combination of more or less random home range exploitation and planned hunting directed toward known places where prey are numerous. In Scotland, Balharry (1993) studied home range size of pine martens in two landscapes. One area was characterized by steep mountains and narrow glens, in which the vegetation was dominated by moorland, low-graded arable grounds and patches of native woodland, and small blocks of commercial forest. The other area consisted of fertile flats, bounded by gentle slopes with a mixture of native woodland and commercial plantations. In the poorer area, the mean home range sizes for four adult males and four adult females were 2,363 ha and 833 ha, respectively. In the fertile area, mean home range sizes were 628 ha for five adult males and 357 ha for six females. Prey availability was not reported. Stone martens living in rural and urban areas can take advantage of human leavings, and in the wintertime they will use bird-feeding tables. Individuals living in less productive areas inhabited by people were found to have larger home ranges than those living in more productive rural areas (Herrmann 1987; Skirnisson 1986).

Influence of Competition

Competition from neighbors of the same sex influences the size of the home range of resident martens. For example, S. Broekhuizen and G. J. D. M. Müskens (unpub. data) radio-tracked three *M. foina* females in the Dutch town of Nijmegen. In the autumn of 1986, an unoccupied area next to the home range of an old female (F1) became occupied by one of her daughters (F2) born that year. Most likely the area had been vacant for some time. Figure 6.2a shows the 90% minimum convex polygons (mcps) of the home ranges of F1 and F2 from September 1986 through April 1987. In early May 1987, F2 died. Within one week, F1 added F2's home range to her own (Fig. 6.2b). On 12 October, a young female (F3) born in 1987 settled in the western part of F1's home range (Fig. 6.2c). In the following weeks, both females used the area, but over time F1 avoided F3's home range. By February and March 1988, F1 had concentrated her activity in the original home range of F2 (likely one of the best marten habitats in the study area), while F3 had taken over the best part of the former home range of F1 (Fig. 6.2d). During this period, F1's home range, measured as the 100% mcp, varied from 7.6 to 140.0 ha; the 90% mcp varied from 5.7 to 61.3 ha. The presence and absence of neighboring females probably was a major cause of this variability.

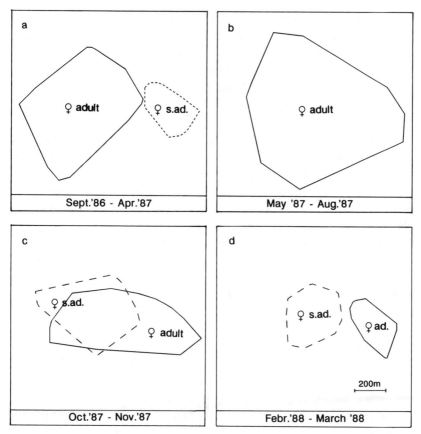

Fig. 6.2. Home range (minimum convex polygon including 90% of radio-fixed locations) of an adult stone marten (solid line) during four intervals from September 1986 to March 1988. (a) From September 1986 to April 1987, a subadult daughter (dotted line) lived contiguously. (b) The daughter died in April 1987. (c) An unrelated subadult female (dashed line) settled in a broadly overlapping home range in October 1987. (d) The two females achieved nonoverlapping ranges by February 1988.

Influence of Seasons

Data on seasonal influences on home range size are conflicting for *M. martes*. Storch (1988) reported that the home ranges of two females in Sweden were about 24% and 17% smaller in summer than in winter (250 vs. 330 ha for female 1, 430 vs. 520 ha for female 2). Marchesi (1989) found that home range size varied seasonally for a male and a female pine marten in the Swiss Jura Mountains, with the largest home ranges during winter. The female's home range was smallest in autumn, the male's in autumn and spring. Schropfer et al.

(1989) conducted a two-year radio-tracking study of a male and a female pine marten in Germany. They found that the home range of the female was smaller in both winters than in the previous summers, although her home range decreased overall. The home range of the adult male was at least 60% smaller in both winters than in the summers. Balharry (1993) also found that males and females covered a smaller proportion of their home range in winter than in summer.

Skirnisson (1986) described seasonal variation in the home range size of two adult *M. foina* females. Both females had the largest home ranges in late summer and the smallest, probably as a consequence of having young, from March to July. Herrmann (1987) measured the smallest home range movements of a male in November and December and the largest in June to July (the latter being the mating season).

S. Broekhuizen and G. J. D. M. Müskens (unpub. data) found that radio-tracking data from five successive nights gave a realistic impression of the size of stone marten home ranges, with the results being similar to Balharry's (1993) findings with Scottish pine martens. Cumulative home range (mcp 100%) stabilized after about five weeks (Fig. 6.3) when animals were located at 15-minute intervals one night per week. Figure 6.4 shows the average home range size (mcp 100% and 90%) of two males sharing the same area: one born in 1987 and the other before 1986. The older male's home range was smallest during winter and largest in late spring and early summer. The home range of the younger (but full-grown) male was largest also in late spring and early summer but was smallest in September to October.

Minimum home range size for an old female was reached in winter (Fig. 6.5). Her maximum home range in 1987 (when she did have a litter) was reached in late spring and early summer. In 1988, her spring home range initially was small but expanded rapidly, perhaps following the death of her litter. Figure 6.6 shows data for a female likely born in 1986. Although she had no litter during the years she was tracked, the periods of minimum and maximum home range size varied; her minimum was reached in winter in 1988 and in late summer and autumn in 1989.

Social Influences

Space use must be interpreted in light of current knowledge of the social structure of the population. For example, in the Dutch stone marten study area in Nijmegen, during the period May to July 1988, the home range of an

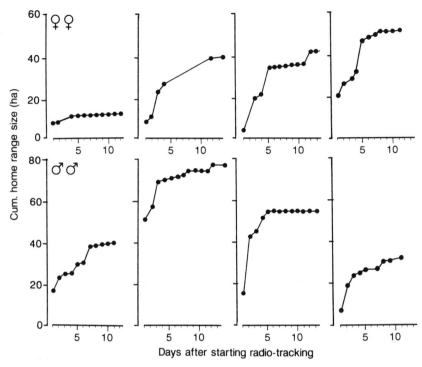

Fig. 6.3. Cumulative home range size in hectares for four female and four male *Martes foina* in Nijmegen, The Netherlands. Dots represent successive nights of radio-tracking and show the minimum convex polygon including 100% of locations.

adult male encompassed the home ranges of three females: two adults born before 1986 and one born in 1987 (Fig. 6.7a). Although the male visited the home ranges of all three females in May (Fig. 6.7b), he spent the daytime in shelters used by one of the two adult females (♀1). That female had given birth to a young, and several times the male was observed to share their day hide. During most of June, the male did not visit this female's home range (Fig. 6.7c). He still visited the two other females and in the daytime rested mostly in shelters used by the second adult female, who had a litter in 1989 but not in 1988. In The Netherlands, the mating season starts in June. It is likely that the male joined the first adult female in order to copulate. By the end of June, the male had lost interest in the day hides of the second female and returned to that of the female with young (Fig. 6.7d). She died the next winter and was not pregnant.

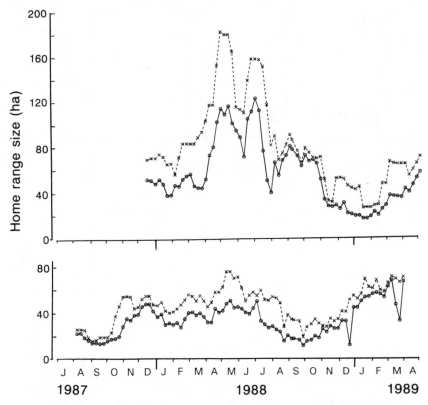

Fig. 6.4. Home range size in hectares (minimum convex polygon) of an adult male *Martes foina* (above) and a juvenile/subadult male (below) that shared the same general area near Nijmegen, The Netherlands, from July 1987 to April 1989. Each point represents a five-night running average. Solid lines are estimates based on 90% of the radio-fixes; dashed lines include 100% of fixes.

How Are Marten Populations Organized?

Exclusivity of Home Ranges

The territorial and home range spatial dynamics of American and European martens follow the general mustelid pattern (Powell 1979), with strong intra-sexual male territoriality overlapping intrasexual territorial female home ranges (Balharry 1993; Clark 1984; Clark et al. 1987; Herrmann 1987). Strick-land et al. (1982) and Strickland and Douglass (1987) reported that immature American martens as well as members of the opposite sex were tolerated, but animals of the same sex exhibited spatial and/or temporal separation (Bis-sonette et al. 1988; Buskirk 1983; Francis and Stephenson 1972; Hawley and

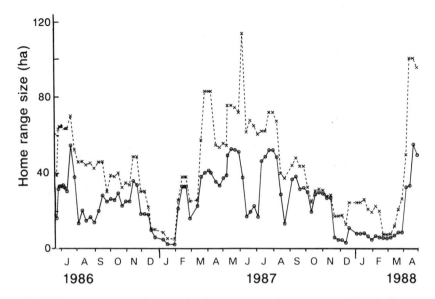

Fig. 6.5. Home range size in hectares (minimum convex polygon) for an old female *Martes foina* living on the outskirts of Nijmegen, The Netherlands, from June 1986 to April 1988. Each point represents a five-night running average. Solid lines are estimates based on 90% of the radio-fixes; dashed lines include 100% of fixes.

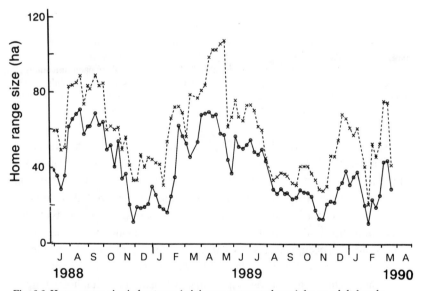

Fig. 6.6. Home range size in hectares (minimum convex polygon) for an adult female (probably born in 1986) *Martes foina* living on the outskirts of Nijmegen, The Netherlands, from June 1988 to March 1990. Solid lines are estimates based on 90% of the radio-fixes; dashed lines include 100% of fixes.

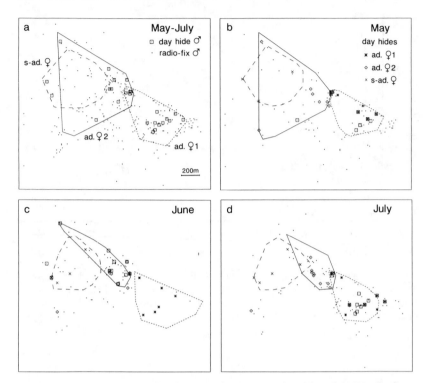

Fig. 6.7. Home ranges of three females (minimum convex polygon based on 90% of radio-fixes) and one male *Martes foina* during May to July 1988 (a) and for each of the three months separately (b-d). Dots show individual radio-fixes for the male in each time interval. Female ranges are outlined with a dotted line for adult ♀1, a solid line for adult ♀2, and a dashed line for a subadult ♀. The daytime rest sites (hides) of the male are shown in part a; female rest sites as well as those for the male are shown in parts b-d.

Newby 1957; Herman and Fuller 1974; Major 1979). Some exceptions seem to occur. Hawley and Newby (1957) reported the almost complete overlap of two adult male home ranges. They did not indicate whether temporal separation was evident or whether a father-son relationship was involved. A male in its second or third year is considered adult even if not sexually active. Balharry (1993) radio-tracked two nonbreeding male European pine martens less than three years old occupying ranges entirely within the ranges of males over three years of age. Clark and Campbell (1979) reported that ranges of female American martens tended to overlap. Balharry (1993) found that home ranges of at least nine of 10 female pine martens two years of age or older were within the home range of a single male. Female territories within the home range of the same male showed small but real overlap, but no females from ranges of different adult males were recorded in the same area. Herrmann (1987) and Müskens et al. (1989) noted some overlap in female stone marten ranges.

Local Movements

The degree of movement throughout home ranges and territories appears to vary with residency status, age, and resource availability. Hawley and Newby (1957) reported that 47 of 85 American martens captured in their study area were transients, eight were temporary residents remaining in the area less than three months, and the other 30 were long-term residents. Residents as well as transients were involved in the fluctuations of composition, although they moved less.

American marten densities appear to fluctuate during the year (Clark and Campbell 1979), apparently as a result of movement patterns as well as dispersal events. Periods of breeding during summer (Markley and Bassett 1942) and of dispersal during late summer to early fall (Archibald and Jessup 1984; Francis and Stephenson 1972) and in early spring (Archibald and Jessup 1984) appear to be times of higher movement (Thompson 1986; Zielinski et al. 1983).

Lensink et al. (1955) reported that movements of martens in Alaska were related to abundance of food resources. Bissonette et al. (1988) and Fredrickson (1990) documented very high movement rates of martens in Newfoundland based on a severe decline in the population of *Microtus pennsylvanicus,* their primary prey choice. Thompson and Colgan (1990) reported lower population densities, larger home ranges, dispersal of adult former residents, and cannibalism as a result of food shortages. Miller et al. (1955) reported American martens covering large areas in pursuit of hares (*Lepus americanus*).

Nyholm (1970) distinguished between resident ("local") and "bounding" European pine martens that moved through the terrain in northern and western Finland. The hunting trips of the local martens varied from 8.5 to 10 km in a night, whereas bounding martens moved 16–65 km in a night. Pulliainen (1981b) also documented both resident and transient pine martens in Finnish Lapland.

Adult pine martens in Scotland occupied the same ranges over a two-year study period, but the proportion of the range used per night varied seasonally (Balharry 1993). In spring and summer, males and females covered a larger proportion of their range per night than in winter. S. Broekhuizen and G. J. D. M. Müskens (unpub. data) found similar results in stone marten. Figure 6.8a shows the areas covered (mcp) per night by two females; Figure 6.8b shows the areas occupied by two males (adult and subadult). Archibald and Jessup (1984) suggested that adult female American martens expand their

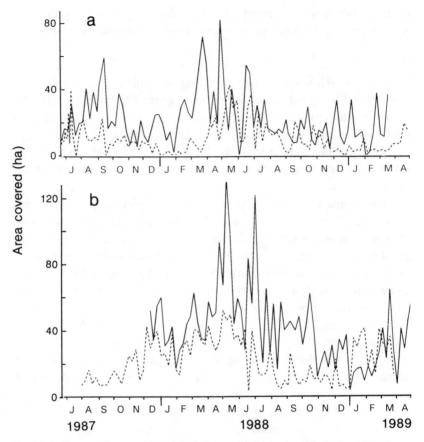

Fig. 6.8. Area covered per tracking night (minimum convex polygon including 100% of radio-fixes) by two adult female *Martes foina* (a) and one adult (solid line) and one subadult (broken line) male (b), from July 1987 to April 1989.

home ranges during estrus to sizes approaching those characteristic of male home ranges. Balharry (1993) suggested that males would increase their defense of females when the females were in estrus. Males defending their breeding rights would be expected to adopt activity patterns different from those of day-to-day foraging.

Dispersal

Juvenile females are known to disperse in the period from late in their first summer until winter (*M. americana:* Brassard and Bernard 1939; *M. martes:* Krott 1973; Meldžiūnaite 1957; Schmidt 1943; *M. foina:* Broekhuizen et al.

1989; Herrmann 1987; Schmidt 1943; Skirnisson 1986). Most authors mention the same period for juvenile male dispersal. Herrmann (1987) reported two young *M. foina* males dispersing only 12 and 15 months after birth. Broekhuizen et al. (1989) describe the case of a young male stone marten, born in spring 1987 in an urban habitat, that remained within the home range of a resident adult male until autumn 1988 (Fig. 6.9a). The resident male was already present in that area in summer 1986 and may have been the parent. Both home ranges overlapped considerably (Fig. 6.9b) until September 1989, when the old male was killed by traffic. During the summer of 1989, the young male's testes were well developed and the animal was obviously sexually mature, but apparently not sexually active. After the old male's home range was taken over by an immigrating male, the young male's home range became strictly separated from that of the new male. Hawley and Newby (1957) reported the almost complete overlap of the home ranges of two adult male *M. americana*. Perhaps intrasexual territoriality is related as much to sexual activity as to age. Heptner et al. (1974) suggested that the time of dispersal for *M. martes* depends on the abundance of food.

Discussion

Until relatively recently, little attention has been paid to scale effects. Biologists and managers alike made little distinction among the levels at which data were collected and applied. The tacit assumption was that results from mechanistic and smaller-scale studies (sensu Price 1986) were applicable for management decisions across landscapes. Recent work in landscape ecology (Forman and Godron 1986; Merriam, Chap. 4; Pickett and White 1986; Wiens et al. 1993) and our recent work (Bissonette and Hargis 1995; C. D. Hargis and J. A. Bissonette, unpub. data) demonstrates clearly that scale effects are extremely important and that explicit a priori consideration of scale is imperative before conservation studies are undertaken. It is not sufficient to take only a large-scale perspective; to do so would be to commit the same error that has characterized past studies, namely, the single-scale consideration of ecological phenomena. We advocate a multiscale approach. Martens are an especially appropriate example of a species for which single-scale approaches have given incomplete answers to the question of what structures population dynamics.

   The most obvious landscape characteristic that appears to influence marten numbers, density, and distribution is fragmentation. Because American martens, and to perhaps a lesser extent pine martens (Brainerd 1990), are so closely tied to mature forested habitats, it is reasonable to expect that

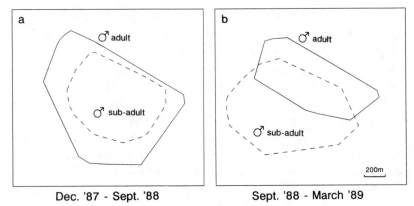

Dec. '87 - Sept. '88   Sept. '88 - March '89

Fig. 6.9. Home ranges (mcp including 90% of radio-fixes) of two *Martes foina* males from December 1987 to March 1989, living on the outskirts of Nijmegen, The Netherlands. One male (solid line) is an adult resident and the other (broken line) is a subadult, possibly a son of the adult.

diminution of either the extent or the age of existing coniferous forest cover would have a negative impact on them. C. D. Hargis and J. A. Bissonette (unpub. data) took a multiscale approach recently to study the effects of fragmentation on marten in the Uinta Mountains of Utah. They reasoned that because martens are core-sensitive habitat specialists (Bissonette et al. 1989), fragmentation of habitat should have at least some impact on population density. At the same time, martens are food-generalist carnivores (Ansorge 1989b; Brainerd 1990; Marchesi 1989; Rzebik-Kowalska 1972; Warner and O'Sullivan 1982) and show opportunistic behavior when foraging. Hargis and Bissonette reasoned that the density of small-mammal prey species and how they are affected by fragmentation also should influence marten densities. In effect, interactions on at least two scales must be operating.

Bissonette and Hargis (1995) and C. D. Hargis and J. A. Bissonette (unpub. data) proposed two models. The first, or *interior model,* postulated that any fragmentation should result in decreased habitat quality. The second, or *partial fragmentation model,* suggested that habitat quality initially rises at low fragmentation levels because of increased abundance of small-mammal species in the early seral stage habitats. Martens did indeed respond negatively to large-scale fragmentation, but their decline was ameliorated by elevated prey densities at much smaller scales. Even in fragmented habitats, marten densities were higher when small mammals were abundant. The spatial arrangement of core habitat, the width of clear-cut patches, and the population response of small-mammal prey along ecotones and edges are important considerations that influence marten population dynamics.

Conclusions

As medium-sized carnivores with large home ranges, martens are suitable for the study of landscape processes. They are predators that are simultaneously habitat specialists and food generalists (Brainerd 1990). *M. americana* and *M. martes* can be used as indicators of older-growth coniferous forest; *M. foina* prefers forest edges and hedgerows. Data on home range core areas and/or marten density can be used to assess habitat quality. However, density data must be interpreted with caution (Van Horne 1983). We can place confidence in relative or absolute densities as indicators of good habitat only if they are coupled with other measures of species well-being—that is, survival, productivity, and recruitment. Further, the differential availability of habitats often precludes a rigorous determination of "preference." For example, if only *n* of *N* habitats are available across a landscape, one cannot legitimately say any habitat is preferred; rather, we must refer to what is selected. Source-sink considerations (Lidicker, Chap. 1; Oksanen and Schneider, Chap. 7; Pulliam 1988) compound the difficulty of habitat assessment. Habitat-specific demographic rates are seldom apparent without close scrutiny. Higher densities of organisms may occur in an area (sink) primarily because of dispersal from a source. The determination of source and sink habitats depends on measurement of habitat-specific birth and death rates and on knowing whether immigration is greater than emigration. In virtually all of the work we report (including most of our own), assumptions are made regarding what is good habitat for martens without providing the demographic information necessary to make a rigorous determination. Habitat core areas suggest relatively good qualities, but they can be modified by social and seasonal factors. Even in very large data sets where these complications can be ignored, investigators may miss the important interactions that really elucidate the causative agents in population responses. Consequently, a multiscale approach is necessary. Such a research strategy has clarified the patterns of marten habitat use and preference, including marten responses to fragmentation.

Summary

In this chapter, we address the spatial habitat patterns of both American and European martens (*Martes* spp.) and discuss differences and similarities in their ecological characteristics. We pose the question of what is good habitat and report evidence from the literature as well as from our own research. We discuss the importance of prey resources and its relation to habitat quality and use. We analyze marten population organization and discuss how competition and seasonal influences affect marten spatial dynamics. Finally, we address the issue of scale as it relates to marten ecology and show that both small-scale effects of prey density and larger-scale habitat fragmentation influence marten densities. We argue that a multiscale research approach is a more effective paradigm for addressing complex resource issues than is an approach that is based on only one scale.

Literature Cited

Amores, F. 1980. Feeding habits of the stone marten, *Martes foina* (Erxleben, 1777), in southwestern Spain. *Saugetierkd. Mitt.* 28:316–22.

Ansorge, H. 1989a. Die Ernährungsökologie des Steinmarders *Martes foina* in den Landschafttypen der Oberlausitz. Pages 473–93 *in* Stubbe, M. (ed.), *Populationsökologie marderartiger Säugetiere.* Wiss. Beitr. 37. Halle-Wittenberg, Germany: Martin Luther University. 647 pp.

Ansorge, H. 1989b. Nahrungsökologische Aspeckte bei Baummarder, Iltis und Hermelin (*Martes martes, Mustela putorius, Mustela erminea*). Pages 494–504 *in* Stubbe, M. (ed.), *Populationsökologie marderartiger Säugetiere.* Wiss. Beitr. 37. Halle-Wittenberg, Germany: Martin Luther University. 647 pp.

Archibald, W. R., and R. H. Jessup. 1984. Population dynamics of the pine marten (*Martes americana*) in the Yukon Territory. Pages 21–97 *in* Olson, R., R. Hastings, and F. Geddes (eds.), *Northern Ecology and Resource Management.* Edmonton: University of Alberta Press. 438 pp.

Balharry, D. 1984. Phoenix of the forests. *British Broadcasting Co. Wildl.* 2:198–203.

Balharry, D. 1993. Social organization in martens: An inflexible system. *Symp. Zool. Soc. London* 65:321–45.

Bateman, M. C. 1986. Winter habitat use, food habits and home range size of the marten, *Martes americana,* in western Newfoundland. *Can. Field-Nat.* 100:58–62.

Bergerud, A. T. 1969. The status of pine marten in Newfoundland. *Can. Field-Nat.* 83:128–31.

Bissonette, J. A., R. J. Fredrickson, and B. J. Tucker. 1988. *The Effects of Forest Harvesting on Marten and Small Mammals in Western Newfoundland.* Report prepared for the Newfoundland and Labrador Wildlife Division and Corner Brook Pulp and Paper, Ltd. Logan: Utah State University. 109 pp.

Bissonette, J. A., R. J. Fredrickson, and B. J. Tucker. 1989. American marten: A case for landscape-level management. *Trans. North Am. Wildl. Nat. Resour. Conf.* 54:89–100.

Bissonette, J. A., and C. D. Hargis. 1995. Linking landscape and smaller scale responses: A multi-scale model. *In* Bissonette, J. A., and P. R. Krausman (eds.), *Proceedings of the First International Wildlife Management Congress: Integrating People and Wildlife for a Sustainable Future (Sept. 19–25, 1993, Costa Rica).* In press.

Brainerd, S. M. 1990. The pine marten (*Martes martes*) and forest fragmentation: A review and general hypothesis. *Trans. Int. Union Game Biol. Congr.* (Trondheim, 1989) 19:421–34.

Brassard, J. A., and R. Bernard. 1939. Observations on breeding and development of marten. *Can. Field-Nat.* 53:15–21.

Broekhuizen, S., M. P. A. Lucas, and G. J. D. M. Müskens. 1989. Behaviour of a young beech marten (*Martes foina* Erxleben 1777) during dispersion. Pages 422–32 *in* Stubbe, M. (ed.), *Populationsökologie marderartiger Säugetiere.* Wiss. Beitr. 37. Halle-Wittenberg, Germany: Martin Luther University. 647 pp.

Buskirk, S. W. 1983. *The Ecology of Marten in Southcentral Alaska.* Ph.D. dissertation. Fairbanks: University of Alaska. 144 pp.

Buskirk, S. W. 1984. Seasonal use of resting sites by marten in south-central Alaska. *J. Wildl. Manage.* 48:950–3.

Buskirk, S. W., S. C. Forrest, M. G. Raphael, and J. J. Harlow. 1989. Winter resting site ecology of marten in the central Rocky Mountains. *J. Wildl. Manage.* 53:191–6.

Buskirk, S. W., H. J. Harlow, and S. C. Forrest. 1987. Studies on the resting site ecology of marten in the central Rocky Mountains. Pages 150–3 *in* Troendle, C. A., M. R. Kaufmann, R. J. Hamre, and R. P. Winokur (eds.), *Management of Subalpine Forests: Building on 50 Years of Research.* Society of American Foresters Publication SAF 87.08, USDA Forest Service General Technical Report RM-149. Silver Creek, Colo.: U.S. Forest Service. 253 pp.

Buskirk, S. W., H. J. Harlow, and S. C. Forrest. 1988. Temperature regulation in American marten (*Martes americana*) in winter. *Natl. Geogr. Res.* 4:208–18.

Buskirk, S. W., and L. L. McDonald. 1989. Analysis of variability in home-range size of the American marten. *J. Wildl. Manage.* 53:997–1004.

Clark, T. W. 1984. Analysis of pine marten population organization and regulatory mechanisms in Jackson Hole, Wyoming. *Natl. Geogr. Soc. Res. Rep.* 1975:131–43.

Clark, T. W., E. Anderson, C. Douglas, and M. Strickland. 1987. *Martes americana.* Mammalian Species 289. Provo, Utah: American Society of Mammalogists. 8 pp.

Clark, T. W., and T. M. Campbell. 1979. Population organization and regulatory mechanisms of pine martens in Grand Teton National Park, Wyoming. Pages 293–5 *in* Linn, R. M. (ed.), *First Conference on Scientific Research in the National Parks,* vol. 1. Trans. Proc. Ser. No. 5. Washington, D.C.: National Park Service. 681 pp.

Clevenger, A. P. 1993. Spring and summer food habitat use of the European pine marten (*Martes martes*) on the island of Minorca, Spain. *J. Zool.* 229:153–61.

Clough, G. C. 1987. Relations of small mammals to forest management in northern Maine. *Can. Field-Nat.* 101:40–8.

Degnn, H. J., and B. Jensen. 1977. Skovmåren (*Martes martes*) i Danmark. *Dan. Vildtuntersøgelser* 29:3–20.

Delibes, M. 1978. Feeding habits of the stone marten, *Martes foina* (Erxleben, 1777), in northern Burgos, Spain. *Z. Saugetierkd.* 43:282–8.

Dolch, D. 1991. Der Baummarder in Ländchen Bellin ausserhalb geschlossener Waldungen. Page 52 *in* Schäfers, G. (ed.), *10. Marder-Kolloquium, Kurzfassungen der Beiträge.* Hamburg: Umweltbehörde Landesforstverwalting. 60 pp.

Forman, R. T. T., and M. Godron. 1986. *Landscape Ecology.* New York: John Wiley & Sons. 619 pp.

Francis, G. R., and A. B. Stephenson. 1972. *Marten Ranges and Food Habits in Algonquin Provincial Park, Ontario.* Research Report (Wildlife) 91. Toronto: Ontario Ministry of Natural Resources. 53 pp.

Fredrickson, R. J. 1990. *The Effects of Disease, Prey Fluctuation, and Clear-cutting on American Marten in Newfoundland, Canada.* M.S. thesis. Logan: Utah State University. 64 pp.

Grupe, G., and H. H. Kruger. 1990. Feeding ecology of the stone and pine marten revealed by element analysis of the skeletons. *Sci. Total Environ.* 90:227–40.

Harestad, A. S., and F. L. Bunnell. 1979. Home range and body weight—A reevaluation. *Ecology* 60:389–402.

Hargis, C. D. 1981. *Winter Habitat Utilization and Food Habits of Pine Martens in Yosemite National Park*. M.S. thesis. Berkeley: University of California. 57 pp.

Hargis, C. D., and D. R. McCullough. 1984. Winter diet and habitat selection of marten in Yosemite National Park. *J. Wildl. Manage.* 48:140–6.

Hawley, V. D., and F. E. Newby. 1957. Marten home ranges and population fluctuations. *J. Mammal.* 38:174–84.

Heptner, V. G., N. P. Naumov, P. B. Jürgenson, A. A. Sludski, A. F. Cirkova, and A. G. Bannikov. 1974. *Die Säugetiere der Sowjetunion. Band II: Seekühe und Raubtiere.* Jena, Germany: Fischer Verlag. 1,006 pp.

Herman, T., and K. Fuller. 1974. Observations of the marten, *Martes americana,* in the McKenzie District, Northwest Territories. *Can. Field-Nat.* 88:501–3.

Herrmann, M. 1987. *Zum Raum-Zeit-System von Steinmarderrüden* (Martes foina, *Erxleben 1777) in unterschiedlichen Lebensräume des südöstlichen Saarlandes.* Bielefeld, Germany: Diplomarbeit Univ. 64 pp.

Holišova, V., and R. Obrtel. 1981. Scat analytical data on the diet of urban stone martens *Martes foina* (Mustelidae, Mammalia). *Folia Zool.* 31:20–30.

Kalpers, J. 1983. Contribution à l'étude eco-éthologique de la fouine *(Martes foina):* Stratégies d'utilisation du domaine vital et des ressources alimentaires. I. Introduction générale et analyse du régime alimentaire. *Cah. Ethologie Appl.* 3:145–63.

Koehler, G. M. 1975. *The Effects of Fire on Marten Distribution and Abundance in the Selway-Bitterroot Wilderness.* M.S. thesis. Moscow: University of Idaho. 26 pp.

Koehler, G. M., and M. G. Hornocker. 1977. Fire effects on marten habitat in the Selway-Bitterroot Wilderness. *J. Wildl. Manage.* 41:500–5.

Krott, P. 1973. Die Fortpflanzung des Edelmarders *(Martes martes* L.) in freier Wildbahn. *Z. Jagdwiss.* 19:113–7.

Lachat Feller, N. 1993. *Eco-éthologie de la Fouine* (Martes foina Erxleben, *1777) dans le Jura Suisse.* Ph.D. thesis. Neuchâtel, Switzerland: Univ. de Neuchâtel. 183 pp.

Lammertsma, D. 1992. Gebruik dagrustplaats door steenmarters. *Zoogdier* 3(4):4–7.

Lensink, C. J., R. O. Skoog, and J. L. Buckley. 1955. Food habits of marten in interior Alaska and their significance. *J. Wildl. Manage.* 19:364–8.

Major, J. T. 1979. *Marten Use of Habitat in a Commercially Clear-cut Forest during Summer.* M.S. thesis. Orono: University of Maine. 48 pp.

Marchesi, P. 1989. *Ecologie et Comportement de la Martre* (Martes martes L.) *dans le Jura Suisse.* Ph.D. thesis. Neuchâtel, Switzerland: Univ. de Neuchâtel. 185 pp.

Marchesi, P., and C. Mermod. 1989. Régime alimentaire de la martre *(Martes martes* L.) dans le Jura suisse (Mammalia: Mustelidae). *Rev. Suisse Zool.* 96:127–46.

Markley, M. H., and C. F. Bassett. 1942. Habits of captive marten. *Am. Midl. Nat.* 28:604–16.

Martell, A. M. 1984. Changes in small mammal communities after fire in northcentral Ontario. *Can. Field-Nat.* 98:223–6.

Martell, A. M., and A. Radvanyi. 1977. Changes in small mammal populations after clearcutting in northern Ontario black spruce forest. *Can. Field-Nat.* 91:41–6.

Martin, S. K. 1987. *The Ecology of Pine Marten* (Martes americana) *at Sagehen Creek, California.* Ph.D. dissertation. Berkeley: University of California. 223 pp.

Martin, S. K., and R. H. Barrett. 1983. The importance of snags to pine marten habitat in the northern Sierra Nevada. Pages 114–6 *in* Davis, J. W., G. A. Goodwin, and R. A. Ockenfels (eds.), *Snag Habitat Management.* Proc. Symp. USDA Forest Service General Technical Report RM-99. Fort Collins, Colo.: Rocky Mountain Forest and Range Experiment Station. 226 pp.

Martin, S. K., and R. H. Barrett. 1991. Resting site selection by marten at Sagehen Creek, California. *Northwest. Nat.* 72:37–42.

Meldžiūnaite, S. 1957. Age determination and age structure of pine martens in Lithuania. *Tr. Akad. Nauk Lit. SSR Ser. B* 3:169–77.

Miller, R. G., R. W. Ritcey, and R. Y. Edwards. 1955. Live-trapping marten in British Columbia. *Murrelet* 36:1–8.

Monthey, R. W., and E. C. Soutiere. 1985. Responses of small mammals to forest harvesting in northern Maine. *Can. Field-Nat.* 99:13–8.

Murie, A. 1961. Some food habits of the marten. *J. Mammal.* 42:512–21.

Müskens, G. J. D. M., L. Meuwissen, and S. Broekhuizen. 1989. Simultaneous use of dayhides in beech martens (*Martes foina* Erxleben, 1777). Pages 409–21 *in* Stubbe, M. (ed.), *Populationsökologie marderartiger Säugetiere.* Wiss. Beitr. 37. Halle-Wittenberg, Germany: Martin Luther University. 647 pp.

Nagorsen, D. W., K. F. Morrison, and J. E. Forsberg. 1989. Winter diet of Vancouver Island marten (*Martes americana*). *Can. J. Zool.* 67:1394–400.

Nesvadbova, J., and J. Zejda. 1984. The pine marten (*Martes martes*) in Bohemia and Moravia. *Folia Zool.* 33:57–64.

Newby, F. E. 1951. *Ecology of the Marten in the Twin Lakes Area, Chelan County, Washington.* M.S. thesis. Pullman: Washington State College. 38 pp.

Nicht, M. 1969. Ein Beitrag zum Vorkommen des Steinmarders, *Martes foina* (Erxleben, 1777) in der Großstadt (Magdeburg). *Z. Jagdwiss.* 15:1–6.

Nyholm, E. S. 1970. On the ecology of the pine marten (*Martes martes*) in eastern and northern Finland. *Suom. Riista* 22:105–18.

O'Sullivan, P. J. 1983. The distribution of the pine marten (*Martes martes*) in the Republic of Ireland. *Mammal Rev.* 13:39–44.

Pelikán, J., and J. Vačkař. 1978. Densities and fluctuation in numbers of red fox, badger and pine marten in the "Bučin" forest. *Folia Zool.* 27:289–303.

Pickett, S. T. A., and P. S. White (eds.). 1986. *The Ecology of Natural Disturbance and Patch Dynamics.* New York: Academic Press. 472 pp.

Powell, R. A. 1979. Mustelid spacing patterns: Variations on a theme by *Mustela. Z. Tierpsychol.* 50:153–65.

Price, M. V. 1986. Structure of desert rodent communities: A critical review of questions and approaches. *Am. Zool.* 26:39–49.

Pulliainen, E. 1981a. Winter habitat selection, home range, and movements of the pine marten (*Martes martes*) in a Finnish Lapland forest. Pages 1068–87 *in* Chapman, J. A., and D. Pursley (eds.), *Worldwide Furbearer Conference Proceedings,* vol. 2. Frostburg, Md. 2,056 pp.

Pulliainen, E. 1981b. A transect survey of small land carnivores and red fox populations on a subarctic fell in Finnish Lapland forest over 13 winters. *Ann. Zool. Fenn.* 18:270–8.

Pulliam, H. R. 1988. Sources, sinks, and population regulation. *Am. Nat.* 132:652–61.

Raine, R. M. 1983. Winter habitat use and responses to snow cover of fishers (*Martes pennanti*) and marten (*Martes americana*) in southeastern Manitoba. *Can. J. Zool.* 61:25–34.

Rasmussen, A. M., and A. B. Madsen. 1985. The diet of the stone marten *Martes foina* in Denmark. *Natura Jutlandica* 21:141–4.

Rasmussen, A. M., A. B. Madsen, T. Asferg, B. Jensen, and M. Roosengaard. 1986. Undersøgelser over husmåren (*Martes foina*) i Danmark. *Dan. Viltundersøgelser* 41:1–39.

Rzebik-Kowalska, B. 1972. Studies on the diet of the carnivores in Poland. *Acta Zool. Cracoviensia* 17:415–506.

Schmidt, F. 1943. *Naturgeschichte des Baum- und des Steinmarders.* Leipzig: Paul Schöps. 528 pp.

Schröpfer, R., W. Biederman, and H. Szczesniak. 1989. Saisoale Aktionraumveränderungen beim Baummarder *Martes martes* L. 1758. Pages 433–42 *in* Stubbe, M. (ed.), *Popula-*

tionsökologie marderartiger Säugetiere. Wiss. Beitr. 37. Halle-Wittenberg, Germany: Martin Luther University. 647 pp.

Sherburne, S. S. 1992. *Marten Use of Subnivean Access Points in Yellowstone National Park, Wyoming.* M.S. thesis. Logan: Utah State University. 37 pp.

Sherburne, S. S., and J. A. Bissonette. 1993. Squirrel middens influence marten (*Martes americana*) use of subnivean access points. *Am. Midl. Nat.* 129:204–7.

Sherburne, S. S., and J. A. Bissonette. 1994. Marten subnivean access point use: Response to subnivean prey levels. *J. Wildl. Manage.* 58:400–5.

Simon, T. L. 1980. *An Ecological Study of the Marten in the Tahoe National Forest, California.* M.S. thesis. Sacramento: California State University. 140 pp.

Skirnisson, K. 1986. Untersuchungen zum Raum-Zeit-System freilebender Steinmarder (*Martes foina* Erxleben, 1777). *Beitr. Wildbiol.* 5:1–200.

Slough, B. G., W. R. Archibald, S. S. Beare, and R. H. Jessup. 1989. Food habits of martens *Martes americana* in the south-central Yukon Territory. *Can. Field-Nat.* 103:18–22.

Soutiere, E. C. 1978. *The Effects of Timber Harvesting on the Marten.* Ph.D. dissertation. Orono: University of Maine. 60 pp.

Soutiere, E. C. 1979. Effects of timber harvesting on marten in Maine. *J. Wildl. Manage.* 43:850–60.

Spencer, W. D. 1987. Seasonal rest site preferences of pine marten in the northern Sierra Nevada. *J. Wildl. Manage.* 51:616–21.

Spencer, W. D., R. H. Barrett, and W. J. Zielinski. 1983. Marten habitat preferences in the northern Sierra Nevada. *J. Wildl. Manage.* 47:1181–6.

Stephenson, R. O. 1984. *The relationship of fire history to furbearer populations and harvest.* Federal Aid in Wildlife Restoration, Final Report, Project W-22-2, Job 7.13R. Juneau: Alaska Department of Fish and Game. 86 pp.

Steventon, J. D., and J. T. Major. 1982. Marten use of habitat in a commercially clear-cut forest. *J. Wildl. Manage.* 46:175–82.

Storch, I. 1988. Zur Raumnutzung von Baummardern. *Z. Jagdwiss.* 34:115–9.

Storch, I., E. Lindström, and J. de Jounge. 1990. Diet and habitat selection of the pine marten in relation to competition with the red fox. *Acta Theriol.* 35:311–20.

Strickland, M. A., and C. W. Douglass. 1987. Marten. Pages 531–46 *in* Novak, M. J., A. Baker, M. E. Obbard, and R. Malloch (eds.), *Wild Furbearer Management and Conservation in North America.* Toronto: Ontario Ministry of Natural Resources. 1,150 pp.

Strickland, M. A., C. W. Douglass, M. Novak, and N. P. Hunziger. 1982. Marten (*Martes americana*). Pages 599–612 *in* Chapman, J. A., and G. A. Feldhamer (eds.), *Wild Mammals of North America.* Baltimore: The Johns Hopkins University Press. 1,147 pp.

Tester, U. 1986. Vergleichende Nahrungsuntersuchung beim Stenmarder *Martes foina* (Erxleben, 1777) in großstädtischem und ländlichem Habitat. *Saugetierkd. Mitt.* 33:37–52.

Thompson, I. D. 1986. *Diet Choice, Hunting Behaviour, Activity Patterns, and Ecological Energetics of Marten.* Ph.D. dissertation. Kingston, Ontario, Canada: Queen's University. 173 pp.

Thompson, I. D. 1987. Numerical responses of martens to a food shortage in northcentral Ontario. *J. Wildl. Manage.* 41:824–35.

Thompson, I. D., and P. W. Colgan. 1990. Prey choice by marten during a decline in prey abundance. *Oecologia* 83:443–51.

Tucker, B. M. 1988. *The Effects of Forest Harvesting on Small Mammals in Western Newfoundland and its Significance to Marten.* M.S. thesis. Logan: Utah State University. 49 pp.

van Bostelen, A. J., and M. D. Verhoog. 1992. Het gebruik van dagrustplaatsen door de steenmarter, *Martes foina* (Erxleben 1777) in een (sub) urbaan milieu. *Rapp. Inst. Bos en Natuuronderzoek.* 42 pp.

Van Horne, B. 1983. Density as a misleading indicator of habitat quality. *J. Wildl. Manage.* 47:893–901.

Warner, P., and P. O'Sullivan. 1982. The food of the pine marten *Martes martes* in Co. Clare. *Trans. Int. Union. Game Biol. Congr.* 14:323–30.

Weckwerth, R. P., and V. D. Hawley. 1962. Marten food habits and population fluctuations in Montana. *J. Wildl. Manage.* 26:55–74.

Wiens, J. A., N. C. Stenseth, B. Van Horne, and R. A. Ims. 1993. Ecological mechanisms and landscape ecology. *Oikos* 66:369–80.

Worthen, G. L., and D. L. Kilgore. 1981. Metabolic rate of pine marten in relation to air temperature. *J. Mammal.* 62:624–8.

Wynne, K. M. 1981. *Summer Home Range Use by Adult Marten in Northwestern Maine.* M.S. thesis. Orono: University of Maine. 19 pp.

Zielinski, W. J., W. D. Spencer, and R. H. Barrett. 1983. Relationship between food habits and activity patterns of pine martens. *J. Mammal.* 64:387–96.

7

The Influence of Habitat Heterogeneity on Predator-Prey Dynamics

Tarja Oksanen and Michael Schneider

To predict the impacts of environmental changes on different species, we must know the relative importance of different population-regulating mechanisms. General ideas on how different processes interact in nature can help us in designing field studies. It is useful to express such ideas explicitly as mathematical or graphical models so that we can make testable predictions. General models are often transferable among ecosystems and can help us in planning conservation policies and in finding realistic answers to management questions.

During the last decade, our research team (see L. Oksanen et al. 1995 for references) has worked on the theory of exploitation ecosystems (Fretwell 1977, analyzed by L. Oksanen et al. 1981) as the main hypothesis for studying interactions among microtine rodents, their resources, and predators in different kinds of habitat complexes. This theory proposes that central aspects of population dynamics can be predicted if we know the trophic position of the species and have a rough idea of the primary productivity of their habitat.

Many of our results have corroborated the predictions of the theory (L. Oksanen et al. 1995). In the course of our studies, however, it has become more and more apparent that the theory in its original form has limited applicability to landscapes with small-scale or mesoscale habitat heterogeneity (resource or habitat patchiness, sensu Ostfeld 1992), which creates interactions across boundaries of different habitat types (see Lidicker, Chap. 1). The theory was thus revised by integrating it with models of habitat selection (T. Oksanen 1990a, T. Oksanen et al. 1992a). In this chapter, we review the modifications that address habitat heterogeneity, describe this development in the light of data obtained from a complex of low arctic tundra and birch brushwoods, and discuss the conservation implications of the modified theory.

Table 7.1. The regulatory mechanism on each trophic level as determined by the length of the food chain in a given system (according to the theory of exploitation ecosystems)

Trophic level	Length of the food chain		
	One link	Two links	Three links
Carnivores	–	–	Competition
Herbivores	–	Competition	Predation
Plants	Competition	Herbivory	Competition

Synopsis of the Modified Theory of Exploitation Ecosystems

According to the theory of exploitation ecosystems (Fretwell 1977; L. Oksanen et al. 1981), population dynamics at different trophic levels depend on the primary productivity of the ecosystem (Table 7.1). In productive terrestrial ecosystems, we expect to find interactions among organisms at three trophic levels: carnivores, herbivores, and plants. Carnivores are predicted to be resource-limited, and we expect their communities to be organized by classical resource competition. Consequently, grazers are regulated by predation, and grazer communities can be viewed as structured by apparent competition (sensu Holt 1977). Grazing pressure is therefore light, and plant communities are again organized by resource competition.

This idea was proposed previously by Hairston et al. (1960) as a generalization applicable to all terrestrial ecosystems and has been recently defended by Hairston and Hairston (1993). Fretwell (1977), however, regarded the original hypothesis as inapplicable to the short-grass plains of western Kansas, where historical evidence suggested that bison (*Bison bison*) had been resource-limited (compare to Sinclair 1977) and where vegetation had always consisted of grazing-tolerant graminoids (compare to McNaughton 1979). To explain the difference between the arid plains and more humid areas with obviously competition-structured vegetation, Fretwell (1977) argued that less productive terrestrial ecosystems are characterized by two-link dynamics: carnivores are relegated to the role of carrion feeders or utilizers of temporary outbreaks of resource-limited grazers. In such circumstances, plants are under intense grazing pressure and exhibit apparent (indirect) competition (Holt 1977). Thus Fretwell's basic idea was that the effective length of the trophic chain depends on primary productivity. Fretwell extended this idea further by proposing that still less productive ecosystems are characterized by one-link dynamics: mobile grazers are absent, and the scanty vegetation is organized by preemptive competition for the few sites where plant life is possible (Fretwell 1987; L. Oksanen 1980).

L. Oksanen et al. (1981) analyzed this framework in terms of three-dimensional exploitation models (Rosenzweig 1973). They found that the hypothesis was logically consistent and derived predictions on the relation between plant biomass and primary productivity. By comparing these predictions to available data, they inferred that the relevant productivity thresholds are about 700 $g/m^2/yr$. Three-link dynamics should prevail in forest habitats (including their successional stages), in tallgrass prairies, and on low arctic willow scrublands. Two-link dynamics should be found in steppe, semidesert, and tundra areas, while polar deserts and high-alpine blockfields should be characterized by one-link dynamics (L. Oksanen et al. 1995).

The theory as formulated by Fretwell (1977) and L. Oksanen et al. (1981) tacitly assumes that habitats with the same productivity cover such large areas that movements of consumers among different habitats do not have a significant impact on local dynamics. In reality, however, organisms often encounter a complex of habitats with very different productivities. Forest biomes harbor patches of relatively barren habitat (e.g., bogs, rock outcrops, habitats with sandy and nutrient-poor soils). Conversely, arid landscapes contain patches of productive forest in contact with running water. Even seemingly homogeneous tundra landscapes are characterized by an order of magnitude of variation in primary productivity due to local differences in moisture, nutrient availability, and length of the growing season (Wielgolaski 1975).

The problems caused by spatial heterogeneity were demonstrated in a study on the population dynamics of microtine rodents in a north boreal landscape consisting of mesic taiga, pine heaths, bogs, and patches of tundra on hilltops (Henttonen et al. 1987; T. Oksanen 1990b), in which one of the ideas tested was the hypothesis of exploitation ecosystems. The results were clearly inconsistent with the predictions of the theory. Predators were present to some extent even in the most barren habitats and clearly not only as transients; for instance, tracks of least weasel (*Mustela nivalis*) were observed on the highest ridges, where the weasel had attempted to prey on snow buntings (*Plectrophenax nivalis*). Predator activity was indeed highest in the most productive habitats, but microtine dynamics did not differ qualitatively in productive habitats and barren ones (Henttonen et al. 1987; T. Oksanen and H. Henttonen, unpub. data). Regardless of primary productivity, the final decline in microtine rodents appeared to be caused by predation.

A basic problem with the original theory is that it does not take into account the possibility that predators supported by productive habitats may influence the dynamics of prey in less productive parts of the landscape (Holt 1984, 1985). To remedy this shortcoming, the theory was modified to include consideration of the impacts of habitat selection (T. Oksanen 1990a; T. Oksanen et al. 1995) and optimal patch use (T. Oksanen et al. 1992a) on the relationship between predator-prey dynamics and primary productivity. Structurally, these models presuppose that the optimal (more productive) habitat occurs as patches embedded in a matrix of suboptimal (less productive) habitat. However, unlike in standard patch use models (Charnov 1976) and some models of habitat selection (Holt 1985), the matrix is not empty. It is assumed to provide resources basically similar to those in the optimal habitat but at such low densities that the consumer population could not persist there in the long run (Pulliam 1988).

First, let us assume that habitat patches are larger than the home ranges of top predators (habitat patchiness, sensu Ostfeld 1992) (Fig. 7.1, model A) and that the habitat choice of top consumers conforms to the *ideal free* habitat selection model of Fretwell and Lucas (1970) and Fretwell (1972). In this case, at equilibrium, habitats that do not support at least a zero growth rate (top right square of Fig. 7.2, model A) for the predator population will not be used. The population densities of predators equilibrate across habitat gradients so that the negative impacts of crowding balance the differences in the intrinsic quality of habitats. Thus, fitnesses of predators are equal across the inhabited habitat gradient. Assuming laissez-faire predation (i.e., the only significant interactions among predators are indirect and express themselves as depletion of shared resources), the constancy of consumer fitness across habitat gradients implies that all habitats with any consumers at all will have the same resource density (Fig. 7.2, model B) (T. Oksanen et al. 1995). This is exactly the prediction Rosenzweig (1971) and L. Oksanen et al. (1981) generated by assuming infinitely large habitat patches with self-contained dynamics.

With ideal free habitat choice, spatial heterogeneity thus results in equilibrium patterns of predator and resource densities similar to those generated by models presupposing a closed system without exchanges between neighboring habitats with different primary productivities. The same threshold productivity for supporting a predator population is predicted in all cases. Because individual predators do not defend territories, they have no reason to settle in a habitat with a lower resource density than in other habitats. Moreover, equal fitness across habitat gradients implies that, at equilibrium, each

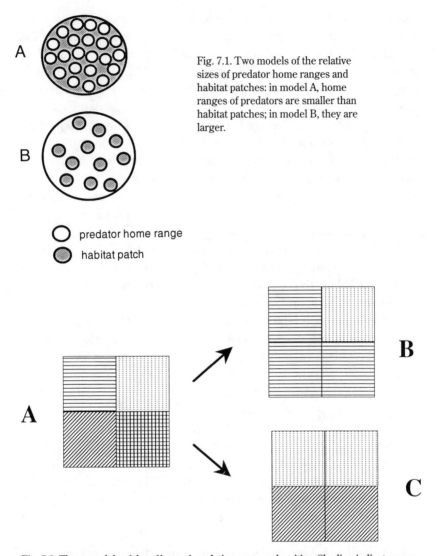

Fig. 7.1. Two models of the relative sizes of predator home ranges and habitat patches: in model A, home ranges of predators are smaller than habitat patches; in model B, they are larger.

○ predator home range
◉ habitat patch

Fig. 7.2. Three models of the effects of predation on prey densities. Shading indicates prey density: very low (dotted), low (horizontal lines), moderate (diagonal lines), or high (cross-hatched). Model A is a system without predators; habitat patches sustain prey population densities corresponding to their productivity. In model B, predators distribute themselves in an "ideal free" manner; prey population densities will be the same in all habitat patches, except in the upper right patch, in which prey density is too low to support predators. In model C, predators show "ideal despotic" behavior, so that density of prey populations correlates positively with productivity. In rich habitats (lower patches), predation pressure is alleviated by negative interactions among predators, while prey populations in poor habitats (upper patches) are affected by spillover predation.

predator replaces itself in all habitats, and there will be no net movement of predators among habitats.

Especially in systems where predators are carnivores, they are often territorial and will defend their essential resources against competitors. When predator density is high enough to make territorial defense economical, intruders must weigh the expected fitness in a given habitat after successful establishment against the risks and fitness costs of losing a territorial contest. The consequence of this extra risk is that habitats with higher intrinsic quality are also characterized by higher average consumer fitness than habitats with lower intrinsic quality. Fretwell (1972) called this model of habitat selection *ideal despotic* (*ideal dominance* in Fretwell and Lucas 1970). It generates source-sink dynamics (Hansson 1977a,b; Holt 1985; Lidicker 1975, Chap. 1; Morris 1994; Pulliam 1988; Pulliam and Danielson 1991) at equilibrium as weak individuals are pushed from better habitats to less productive ones, both intraspecifically and interspecifically.

Quality differences between microhabitats (Pulliam 1988) and stable social hierarchies (Łomnicki 1978) have similar consequences for population dynamics. In such situations, the weakest individuals, which suffer the full impact of crowding, are free to choose among the lowest social ranks or worst microhabitats in each habitat type. Dominant individuals are less severely influenced by crowding, and their fitnesses correlate positively with the intrinsic quality of the habitat. Consequently, intrinsic habitat quality and the average fitness of resident consumers will also be positively correlated.

Whether they are due to despotic or preemptive behavior, differences in the fitness of consumers will, at equilibrium, create a net flux of consumers from more productive to less productive habitats. This flow can be included in Rosenzweig-MacArthur type predator-prey models (Rosenzweig and MacArthur 1963) as a density-dependent emigration term in the predator dynamics of productive habitats and as a density-independent immigration term in unproductive areas. Immigration ensures that there will be predators even in unproductive habitats that cannot support predator populations of their own. This phenomenon has been called *spillover predation* (Holt 1984). If the reproductive output of the prey correlates positively with the productivity of the habitat, then prey populations living in unproductive habitats can actually be more severely depleted by predation than those residing in habitats with high productivity (Fig. 7.2, model C). This result is inconsistent with the model of L. Oksanen et al. (1981), which predicts positive correlation between predation pressure and the productivity of the habitat.

The net flux of predators from productive (optimal) habitat to unproductive (marginal) habitat also enhances the stability of the system by creating direct density-dependence in the productive habitat (for an analytical treatment, see Holt 1984; Łomnicki 1978; Morris 1991; Pulliam 1988; Rosenzweig and MacArthur 1963), which otherwise might be at risk of instability from overexploitation of the prey. The model of T. Oksanen (1990a) assumes that the emigration rate from the productive habitat is independent of the abundance relationships between productive and unproductive habitats. Thus, the model is indirectly connected to Lidicker's (1988) ratio of optimal to marginal patch areas (ROMPA) hypothesis (see also Chap. 1), although the connection has not yet been explicitly analyzed. It seems logical to assume that territorial defense in optimal patches is easiest if marginal habitats cover wide areas and can thus easily absorb large numbers of immigrants, and if the quality of marginal habitats is reasonably high, so that predators are less motivated to make risky attempts to stay in optimal habitats. We thus expect the stabilizing impacts of net dispersal from optimal to marginal habitats to be maximal when ROMPA is low, as proposed by Lidicker (Chap. 1) and Gaines et al. (1991), and when the quality of marginal habitats is relatively high (but not high enough for positive population growth [Ostfeld 1992]). Predation pressure experienced by prey in marginal habitats should increase with increased ROMPA (compare to T. Oksanen 1990b).

The spatial scale of habitat heterogeneity may be such that several habitat types are included in the home range or territory of an individual predator (resource patchiness, sensu Ostfeld 1992) (Fig. 7.1, model B). The consequences of the exploitation of such landscapes can be analyzed on the basis of the Charnov model of optimal patch use (Charnov 1976). The theory predicts that patches will be left when the gain rate within the patch falls below the mean gain rate in the landscape as a whole. If the predator population is at equilibrium with the resource, this average gain rate must be equal to the rate that the predator population needs in order to have zero growth rate (T. Oksanen et al. 1992a). The original model of optimal patch use thus predicts that only those habitats that would support predators if they covered infinitely large areas (top left and bottom squares in Fig. 7.3, model A) will be exploited when they occur as small patches. Moreover, the model predicts that residence times in productive patches will be longer than residence times in less productive ones and that resource densities will be reduced to exactly the same level in all exploited patches (Fig. 7.3, model B), although in the absence of predators, prey densities in exploited patches (Fig. 7.3, model C) converge toward the levels observed in systems without predators (Fig. 7.3, model A). These predictions are quite similar to those of L. Oksanen et

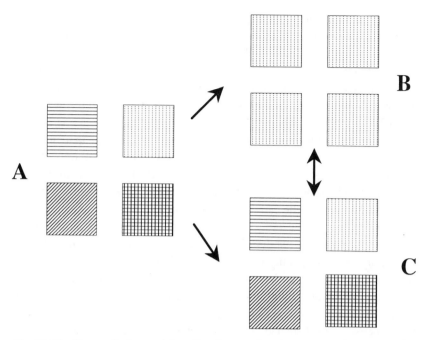

Fig. 7.3. The Charnov-Parker model predicts that a predator's use of a patchy prey population should create local prey fluctuations where minimum densities are equal across habitat gradients, whereas maximum densities correspond to the productivity of the habitat. Patch shading indicates prey density: very low (dotted), low (horizontal lines), moderate (diagonal lines), or high (cross-hatched). In model A, predators are absent and prey densities correspond to patch productivities. In model B, predators are present or have just left the patches. In model C, predators have just arrived at the patches.

al. (1981). Provided, therefore, patch use is optimal in the sense of the classical Charnov model, the model of L. Oksanen et al. (1981) should be fairly realistic even for landscapes with small-scale habitat heterogeneity.

The classical patch use model, however, has one critical limitation. It presupposes that prey are infinitely small and infinitely numerous so that the accumulated yield is a continuous function of the time spent in the patch. Normally, prey items are big enough to create clearly stepwise yield curves, which can be approximated only roughly by continuous functions. We must, therefore, also think about the optimal behavior of a predator that happens to encounter a prey in a normally rejected habitat, which the predator has to pass through on its way between the patches it prefers to exploit. In this situation, we can apply the classical prey choice model (see Stephens and Krebs 1986 for references), which states that a prey should be taken if the ratio of energy gain to handling time exceeds the gain obtained from alternative prey divided by search time plus handling time. Because the choice situation for a transient predator is somewhat different (it is not choosing between two prey types but between predation and transit behavior), the classical rule has to be modified as follows: exploit a haphazardly encountered prey if the ratio of energy gain to handling time exceeds the average gain rate in the landscape as a whole (T. Oksanen et al. 1992a). More exactly, the time lost in shifting from transit to hunting and back to transit should be added to handling time, and the energy costs of these behavioral changes should be subtracted from the yield of the prey. Opportunistic predation by a transient predator is especially likely to be profitable in systems where individual prey items provide a substantial fraction of the daily energy needs of a predator.

In other words, haphazardly encountered prey are treated as if they were a small patch of a better habitat. When a prey has already been encountered, search time is zero. Haphazardly encountered prey are thus superior to the very best patches in the landscape, unless the prey types encountered in transitory habitat are characterized by very low ratios of energy gain to handling time or unless the shift from transit behavior to hunting and back to transit consumes substantial amounts of time or energy (Fig. 7.4). We have thus derived another mechanism for spillover predation. This reasoning implies that prey residing in habitats too barren to support a predator population will nevertheless be exploited.

Differences among habitat types in predation pressure will depend on the fraction of the landscape each type represents. The model of L. Oksanen et al. (1981) passes as a fair approximation of the dynamics in landscapes where unproductive habitats predominate but productive habitats still make up a substantial fraction of the area. In other kinds of habitat configurations, spatial patterns in predation pressure are not even approximately similar to the predictions of L. Oksanen et al. (1981). If unproductive habitats prevail overwhelmingly, the landscape is not likely to support any predators at all. Conversely, small barren areas in generally productive landscapes can be subjected to such intense spillover predation that prey is eliminated from these habitats (Fig. 7.5) (T. Oksanen et al. 1992a).

If the scale of habitat variation is very small, then carnivores will be unable to distinguish between different habitats and will exploit the complex in a fine-grained manner (sensu MacArthur and Levins 1964). Thus, in the carnivore's view, the system has only a single prey population, and the equilibrium will be for the landscape as a whole. Prey, however, have different carrying capacities in the two habitat types and therefore view the landscape as

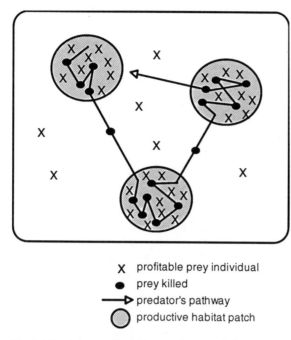

X   profitable prey individual
●   prey killed
──▷ predator's pathway
◯   productive habitat patch

Fig. 7.4. Predation "spillover" from rich to poor habitat patches. An optimally foraging predator will consume profitable prey items not only within productive habitat patches, but also when it encounters them in barren habitats while traveling between productive patches.

coarse-grained. Consequently, predation pressure may well be heavier on the more barren patches than on the productive ones (T. Oksanen et al. 1992a).

To summarize, if habitat selection of predators is despotic or preemptive, if opportunistic predation is profitable, or if predation is fine-grained, then both productive and barren habitats will get some predation, although activity levels will normally be higher in the productive habitat. Moreover, the intensity of spillover predation will depend on the abundance relationships between the two habitat types. The two mechanisms creating spillover predation (despotic behavior and opportunistic predation by transients) share a fundamental similarity. One habitat yields an energy surplus, which subsidizes exploitation in habitats that by themselves would not support exploiters. The larger the fraction of the area that consists of productive habitats, the more intense the predicted spillover exploitation.

Although we have focused so far on spillover carnivory, the same mechanism should also operate for grazer-plant interactions at the transition be-

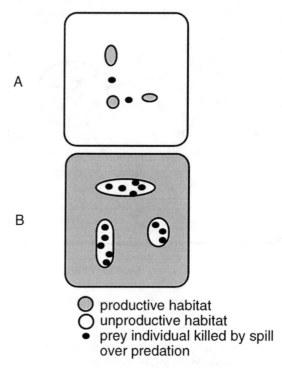

○ productive habitat
○ unproductive habitat
● prey individual killed by spill
over predation

Fig. 7.5. Differences among habitat types in predation
pressure depend on the fraction of the entire landscape
each type covers. When unproductive habitat prevails (A),
the few productive patches support no or only a few predators, and spillover predation is slight relative to the entire
prey population in the unproductive habitat. When productive habitat prevails (B), supporting a high predator population, prey mortality is high in the patches of unproductive habitat.

tween two-link and one-link ecosystems (T. Oksanen 1990a; T. Oksanen et al.
1992a, 1995; see also discussion of the ROMPA hypothesis in Lidicker, Chap.
1, and Hansson, Chap. 2). The original theory of exploitation ecosystems
should apply to landscapes where all habitats are on the same side of the critical productivity thresholds determining the number of trophic levels. In
landscapes spanning both sides of the threshold, the original theory should
be fairly realistic for the prevailing habitat type of the landscape and for rare
habitats with higher productivity, provided that the patches are large or
occur as fairly continuous strands (which keeps travel distances reasonably
short). However, dynamics within minor habitats that are essentially less

productive than the landscape as a whole should be entirely different from the predictions of the original theory.

A Case Study on the Habitat Use of Small Mustelids

Since 1986 we have used snow-tracking to study the habitat use of small mustelids within a 16.1 km² area consisting of various tundra habitats and patches of mountain birch woodland (Fig. 7.6) at the boundary between the Fennoscandian Shield and the Scandinavian mountain range, the Scandes, at Joatka, Finnmark, Norway (L. Oksanen and T. Oksanen 1981, 1992; T. Oksanen et al. 1992b). The area was divided into four subareas (Fig. 7.7). The "highland" subarea (4.2 km²) lies above the thrust cliff of the Scandes at altitudes of 550–672 m. It is entirely treeless and by and large consists of relatively barren tundra habitats (heaths, bogs, snow beds, fell-fields). Relatively productive meadow habitats cover 0.8% of the highland.

The "slope" subarea (2.9 km², altitude 385–550 m) is by far the most productive because of a combination of favorable factors: southern exposure, nutrient-rich bedrock, and abundant seepage areas and spring-fed creeks. Productive habitats (dwarf birch [*Betula nana*] and willow [*Salix*] thickets, herb-rich birch [*B. pubescens*] woodlands, herb-rich mires) cover 82% of the area, while 12% consists of relatively unproductive habitats (lichen and crowberry [*Empetrum hermaphroditum*] heaths, open bogs).

The rest of the study area lies on the Fennoscandian Shield and is chiefly occupied by relatively unproductive habitats (lichen heaths and open bogs). In the "lowland" subarea (5.0 km², altitude 385–440 m) immediately below the slope, productive habitats (willow thickets, blueberry [*Vaccinium myrtillus*] heaths, birch woodlands, herb-rich wetlands) are fairly abundant, covering 16% of the land area, compared to only 2.9% in the "divide" subarea (4.1 km², altitude 415–475 m). The difference between the last two subareas is caused by topography. The lowland is a slight depression below the thrust line of the Scandes and is rich in drainage systems following the thrust line running from the Scandes to the depression. The divide consists of typical shield landscape, with gently rolling hills and very slight altitudinal gradients in drainage systems.

In broader geographic terms, the highland continues right to the coastal cliffs and is thus representative of the peninsulas of northernmost Norway. The divide is representative of the Precambrian tundra landscape between the Scandes and the coniferous forests of Finnish and Swedish Lapland. The slope and the lowland are local landscape types that occur at the limit between the Fennoscandian Shield and the Scandes.

Fig. 7.6. View of the study area at Joatka, Finnmark, Norway. (Photo by M. Schneider, taken in July 1991.)

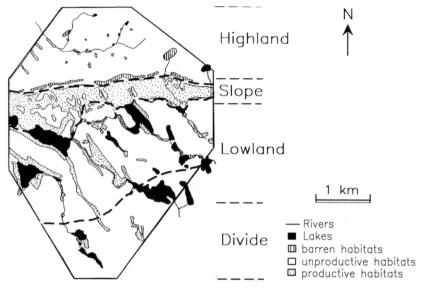

Highland

Slope

Lowland

Divide

N

1 km

— Rivers
■ Lakes
▥ barren habitats
□ unproductive habitats
▤ productive habitats

Fig. 7.7. Map of the study area at Joatka, Finnmark, Norway.

To study the habitat use of small mustelids, we mapped all fresh mustelid tracks within the entire study area during days with calm weather and fresh snow and at least 12 hours after the latest snowfall. A team of two to three field biologists on skis normally did the mapping in early winter (late October to early December). Typical routes favored by mustelids (watercourses, steep slopes [Erlinge 1977]) and certain routes in less preferred areas were checked every tracking day in every subarea; fresh tracks encountered in these areas were followed wherever they went within the boundaries of the study area.

The tracks were then drawn on a topographic map (scale 1:14, 142). To compute an index of stoat (*Mustela erminea*) and weasel (*M. nivalis*) activity in different habitats and subareas, we laid over the map a transparency derived from a vegetation map in the same scale (divided into squares representing 20 by 20 m) and divided the numbers of squares with tracks by the total number of squares representing the habitat type in the subarea in question and by the number of tracking days. To avoid decimal values with many zeros, the indices were multiplied by 500. The indices can thus be interpreted as percentages of the area visited by mustelids during a five-day period (T. Oksanen et al. 1992b).

To identify individuals, we measured the widths, lengths, and depths of the footprints left by the animals and the intervals between 20 consecutive leaps. Tracks were regarded as belonging to the same individual if there was no evidence to the contrary, such as gaps between two sets of tracks, two parallel trails in the same direction, or clear differences in leap lengths or track widths. This method is somewhat conservative, because two individuals of the same species and sex could have been regarded as one if their tracks were intertwined or if they had been active in the same area on different days. Problems with intertwining tracks, however, were limited to the slope subarea and to years with especially many stoats and weasels. In the vast majority of cases, track formations were well separated in space.

Snap-trapping along a transect line, which ran along the longitudinal axis of the study area and intersected all subareas, was used to study the population dynamics of microtine rodents within the area (L. Oksanen and T. Oksanen 1992; P. Ekerholm, pers. comm.). In addition, livetrapping was performed on one big (4 ha) grid and 14 relatively small (0.4–0.6 ha) grids dispersed over all subareas.

The results for 1986–90 have been summarized by T. Oksanen et al. (1992b) and L. Oksanen and T. Oksanen (1992). More recent tracking data are summarized in Table 7.2 and Figure 7.8. The pattern of habitat use has remained the same from 1986 to 1994: highest activities are consistently in the slope subarea. Predator activity in the lowland is lower but still substantial. In the divide, mustelid activity is sporadic, and in the highland, mustelids are practically absent. Within each subarea, mustelid activity is concentrated in the most productive habitats. However, the difference between habitat types is relatively small in the slope area, more pronounced in the lowland, and extremely clear in the divide, where mustelid activity is negligible in barren habitats. Moreover, the activity patterns of the two species differed consistently: stoats showed consistently higher fidelity to the slope and to the most luxuriant habitats there, whereas weasel activity was spread more evenly among subareas.

Small-mammal data until 1990 can be summarized as follows. The vole guild had three cyclic peaks: in 1978–79, 1982–84, and 1987–88. The dominant species was consistently the grey-sided vole (*Clethrionomys rufocanus*), a habitat generalist that occurs both in relatively productive habitats (especially in shallow, hummocky wetlands) and in barren lichen–moss–dwarf shrub tundra. In productive meadows, scrubland, and woodland habitats, redbacked voles (*C. rutilus*), root voles (*Microtus oeconomus*), and field voles (*M. agrestis*) were frequently codominants, especially in the slope subarea (for details, see L. Oksanen and T. Oksanen 1992).

Table 7.2. Activity indices (habitat preferences) of stoats and weasels at Joatka, Finnmark, Norway, during 1991–94

Species, subarea, and habitat	Tracking date		
	Nov. 1991	Nov. 1992	Jan. 1994
Mustela erminea			
Barren habitats			
Highland	–	–	2.25
Unproductive habitats			
Highland	–	1.30	0.10
Slope	15.90	26.25	32.95
Lowland	1.95	8.30	4.95
Divide	0.40	3.40	–
Productive habitats			
Highland	–	–	–
Slope	28.60	69.70	51.30
Lowland	11.75	46.25	32.80
Divide	–	26.60	–
M. nivalis			
Barren habitats			
Highland	–	–	–
Unproductive habitats			
Highland	–	–	1.00
Slope	–	12.90	–
Lowland	–	2.40	1.40
Divide	–	5.65	–
Productive habitats			
Highland	–	–	–
Slope	–	9.15	–
Lowland	–	10.05	3.75
Divide	–	44.35	–

Note: To convert tracking data to activity indices per subarea and habitat, we summed the number of grids (20 × 20 m) visited by each individual, divided this by the total number of grids of the habitat in the subarea, and multiplied by $500/n$, where n is the number of tracking days.

Lemmings (*Lemmus lemmus*), which can be regarded as specialists of barren habitats (Moen et al. 1993; T. Oksanen 1993), had two outbreaks in the highland. The first one, in 1978, occurred during a rise in vole populations and resulted in a wholesale invasion of low-altitude habitats. The second one, in 1988, coincided with the start of a vole decline and was largely restricted to the highland; however, fair numbers of lemmings were also trapped in the divide (Moen et al. 1993; T. Oksanen 1993). Immediately after crashes, a few

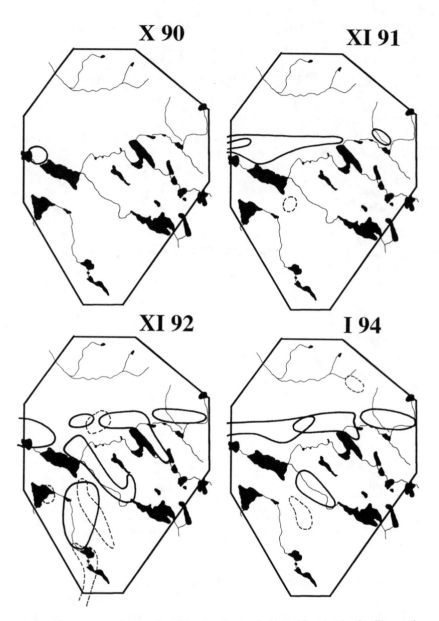

X 90

XI 91

XI 92

I 94

Fig. 7.8. Home ranges of stoats (solid lines) and weasels (dotted lines) at Joatka, Finnmark, Norway, during 1990–94. Lines were drawn convexly around tracks assigned to the same individual but were indented to exclude terrain where the distance to the nearest track was more than 100 m to yield semiconvex home range boundaries. Roman numerals denote months.

lemmings were trapped in the lowland. Otherwise, all lemmings trapped during the periods of low numbers and during the early parts of increases were found in the highland or in meadows and scrublands immediately below the 550 m contour.

The recent trends can be summarized as follows. After a crash in 1988–89, microtines recovered in 1990–91 and since then have reached fairly high autumnal densities. Vernal densities, however, have been relatively low because of substantial winter declines (P. Ekerholm, pers. comm.). A similar syndrome was observed at Pallasjärvi, Finnish Lapland, in the 1980s (Hanski et al. 1993; Henttonen et al. 1987).

These data illustrate how spillover predation can function in practice (Fig. 7.8). The socially subordinate weasels shun areas with high stoat activity and thus end up using more barren parts of the landscape during periods of high stoat numbers (Erlinge and Sandell 1988). This result corresponds to the predictions of the large-patch model of T. Oksanen (1990a). Moreover, both stoats and weasels move in barren habitats, especially where the barren habitats occur in close contact with productive habitats, as predicted by the small-patch model of T. Oksanen et al. (1992a). Empirically, it is often difficult to distinguish between the two mechanisms responsible for spillover predation. Productive habitats often occur as strands or semicontinuous strings of patches, with numerous invaginations of less productive habitat. The borderline between habitat choice and patch use can thus be diffuse in nature.

The relatively modest predator activity found in relatively barren low-altitude areas (the divide, unproductive parts of the lowland) seems nevertheless to suffice to keep local vole fluctuations in phase with the vole cycle typical of productive low-altitude habitats. Moreover, the long-term persistence of lemmings appears to depend critically on the existence of areas with virtually no mammalian predators during any phase of their erratic outbreak-crash sequences. Thus, even relatively modest spillover predation appears to have a major impact on both the numerical dynamics of herbivores and the composition of the grazer community.

Limits of the Applicability of Our Results

Progression from the model of L. Oksanen et al. (1981) to the models of T. Oksanen (1990a) and T. Oksanen et al. (1992a; unpub. data) can be seen as a process whereby spatial structure has been added to classical (Lotka-Volterra-Rosenzweig) predator-prey models (see also Brown and Rosenzweig 1986; Holt 1984, 1985). In these modifications, the landscape is divided into productive patches and a relatively unsuitable habitat, normally assumed to comprise the matrix in which the productive patches are embedded. The key word here is *relatively*. A salient feature of these models is that predators can easily cross the matrix. In the models of Holt (1984) and T. Oksanen et al. (1992a), each territory or home range consists of both productive patches and matrix habitat, whereas Holt (1985) and T. Oksanen (1990a) assume that productive patches are larger than the home ranges or territories of predators. Dynamics on the metapopulation level (Hanski 1991) are trivial, because the unproductive matrix is a population dynamical sink, populated by predators that are highly motivated to fill whatever vacant territories there are in productive patches.

With ideal free habitat selection (Fretwell and Lucas 1970), the matrix habitat has no resident predators (Holt 1984; T. Oksanen et al. 1995). A central premise of the ideal free model, however, is that predators are aware of resource densities in different habitat patches, which means that at least some age class must move freely in the entire landscape and settle preferentially in habitats with the highest availability of resources. If any local predator population is reduced as a result of demographic or environmental stochasticity, the patch will attract settlers. Local extinctions are thus practically impossible under ideal free habitat selection.

When the patches are farther apart and the matrix habitat more hostile, it becomes difficult or at least costly for predators to compare the ambient resource levels in different habitat patches (Morris 1987). Consequently, local dynamics tend to drive down the numbers of prey (or hosts) in patches with many predators, and population growth in different patches need not be in synchrony. Crowley (1981) and Reeve (1988, 1990) have modeled explicitly the predator-prey dynamics within such a system of weakly connected patches. Another approach is to work at the landscape level from the statistical consequences of patchiness, aggregation, and imperfect resource tracking (Comins et al. 1992; Free et al. 1977; Hassell and Pacala 1990).

The approaches discussed above do not explicitly consider the consequences of environmental stochasticity, and local extinctions are dealt with indirectly or not at all. When the matrix habitat is still more hostile and the dispersal distances even longer, some productive patches are likely to be devoid of predators because of stochastic extinctions. Situations of this kind can be analyzed with structured metapopulation models of predator-prey interactions (Diekmann et al. 1989; Gyllenberg and Hanski 1992; Hastings 1977; Kareiva 1990; Sabelis and Diekmann 1988; Sabelis et al. 1991; Taylor 1988, 1990).

The key questions for the choice of approach are thus first, how bad is the matrix? and second, how isolated are the good patches? We suspect that the answers to these questions will vary from system to system. If the matrix habitat can be easily crossed and provides some resources, then the situation should normally correspond to the models of Holt (1984, 1985), T. Oksanen (1990a), and T. Oksanen et al. (1992a). We regard this as a likely situation for vertebrates living in mainland areas.

Implications for Conservation Purposes

The scale of interactions (Wiens et al. 1986) can have important dynamical consequences for both predator and prey, as discussed above. What, then, can we learn from these results for conservation purposes in a time when the space for natural communities is steadily shrinking and many species' habitats are diminished and fragmented (Bright 1993; Jacobs 1988; Saunders et al. 1993)? First, we can no longer view a local community in isolation from the surrounding landscape mosaic (Zonneveld and Forman 1990). As Hobbs (1993a) pointed out, we should examine the "ecosystem effects of fragmentation, as opposed to the biogeographic aspects usually considered." However, we here want to examine not only changes in abiotic factors (Hobbs 1993b), but also changes in biotic interactions.

The influence of spatial heterogeneity on single species and on interactions within trophic levels has been investigated theoretically in great detail (for recent reviews on mammals and birds, see Andrén 1994 and Bright 1993). Despite much empirical work on vertebrate predator-prey systems (e.g., classical studies by Kruuk [1972], Mech [1966], and Schaller [1972]), the landscape view is severely underrepresented. The effects of regional and landscape patterns have been largely ignored until recently (Angelstam et al. 1984, 1985; Bissonette and Broekhuizen, Chap. 6; Doncaster and Krebs 1993; Hansson 1977a; Hansson and Henttonen 1985; Lindström 1989; Pulliainen 1981; Sonerud 1986). In this section, we focus on how carnivore-herbivore dynamics on a landscape scale can be related to questions of species conservation.

The most important changes in species' habitats caused by people include the following: habitats become fragmented; habitat patches decrease in size and number; the distance between habitat patches increases; the distribution of prey animals is altered; the productivity of habitat patches increases; the quality of habitat patches decreases; and the importance of edge effects increases (see also Andrén 1994). Clearly, two or more of these processes often act in concert. All of these changes influence the interactions of predators and their prey in a given landscape and should be important to the population dynamics of both. We now present four scenarios illustrating different landscapes with species reacting in different ways. Illustrative examples are given where possible.

The Community Perspective on Habitat Fragmentation

When a uniform habitat is fragmented, new types of habitat emerge. As fragmentation and alteration proceed, the landscape turns into a mosaic of patches that differ in quality and support habitat for different species. Consequently, the diversity of species is higher in such a heterogeneous landscape than in the homogeneous habitat that formerly existed in the same place. The large number of species encourages the occurrence of generalist predators that can, and do, switch between prey species depending on their relative abundances. Thus, the quality of prey population regulation changes as specialist predators lose their predominant role. For further discussion of this scenario, see Ambuel and Temple (1983), Andrén and Angelstam (1988), Andrén et al. (1985), and Martinsson et al. (1993).

Changes in Area Relationships through Changes in Habitat Patch Size

Imagine a landscape consisting largely of low-productivity two-link communities (plants and herbivores only). The herbivore community is structured by competition, as just a few predators spill over from the small patches with productive three-link systems. The larger the predator population in the productive patches, the more predators disperse into unproductive patches. Now suppose that because of human intervention, the area of the unproductive system diminishes and that of the productive patches increases. Then many more predators will spill over into the unproductive patches. Consequently, the herbivore community becomes structured by apparent competition rather than by classical resource competition; this switch may lead to the disappearance of some species from the landscape or to invasion by others (*predator-mediated coexistence;* see Holt 1977, 1987; T. Oksanen 1993; T. Oksanen et al. 1992a).

If predators, while foraging, temporarily cross the boundaries of their productive habitat patches, they can affect the zone of unproductive habitat adjacent to their prime habitat. Thus, the prey species living in the unproductive area is more influenced by the predator at the edges of its habitat than in the central parts. The larger the edge-to-area ratio of a given unproductive habitat patch, the more detrimental the impact on the prey population as a whole.

Habitat Patches Disappear; Distances between Them Increase

Suppose that habitat patches become smaller than individual home ranges of predators but remain larger than the home ranges of the prey. Each predator would then require several patches and should use them according to opti-mal foraging theory. To maximize net energy intake in the home range as a whole, predators should deplete patches according to the distance between them and the resulting interpatch travel time (Charnov 1976). The greater the interpatch distances, the longer a predator stays in each patch, and the longer the search time, the more the patches become depleted.

The disappearance of habitat patches and the subsequent increase in the distance between patches affect not only the predators but also the prey. The new equilibrium reached is likely to differ from the former one in that fewer predators can survive in the landscape as a whole, and average prey popula-tion densities will be higher in the habitat patches that persist. This second difference arises because as predator travel times and intrapatch search times get longer, the frequency of predators visiting a given patch decreases, so the prey population can temporarily rise to a higher level than before (T. Oksanen et al. 1992a). (Patches that are very remote should be an excep-tion, as visits to them may be so rare that predation may be negligible.) The fewer habitat patches that remain, the fewer predators they can support. Below a threshold number of patches, the predator population will go extinct. The assemblage of former prey species then escapes regulation by preda-tion, and their trophic level will be structured by competition instead, which may lead to the exclusion of some species.

When habitat patches are larger than individual home ranges of predators, the situation should be more favorable. The fact that predators can now per-sist as a metapopulation should enhance stability and thus their survival. Prey populations would occasionally escape regulation by the predator when-ever a local population of predators became extinct, but in general prey would behave as in a three-link system.

Artificial Creation of High-Quality Patches

Even remote natural and seminatural habitats are affected by anthropogenic pollutants, which range from directly toxic substances to a limiting resource. If resources are imported, the productivity of a given habitat patch may in-crease significantly. The quality of biotic interactions will change following an increase in the number of trophic levels, and suddenly, predators will be able to persist in habitats where previously the prey population was too low.

On a large spatial scale, artificial high-quality patches may function as refuges and hence as source populations for predators that spill over into and thus influence surrounding low-quality habitat. In the subarctic regions of Finnish Lapland and Finnmarksvidda in northern Norway, stoats use the garbage dumps at tourist lodges intensely. Stoats persist in such places, even while voles and their predators crash elsewhere (T. Oksanen and M. Schneider, pers. obs.). When such anthropogenic "hot spots" multiply (as is likely in the future), the natural dynamics of predators and prey in the landscape may be severely disturbed (recall the influence of predation on lemming dynamics discussed earlier).

On a small spatial scale, individual predators should use artificial high-quality patches according to the predictions of foraging theory. Many wolves (*Canis lupus*) in Italy visit garbage dumps near human settlements, which represent patches of high resource value to them. They feed on garbage at night and during the daytime retreat to remote mountainous regions, where they escape human interference but where natural prey is scarce. Garbage dumps may help save the Italian wolf populations from extinction until natural prey populations are re-established at sufficiently high levels (Boitani and Ciucci 1993; Zimen 1990a,b).

In many parts of the world, humans feed ungulates during the winter in order to increase survival of popular species for hunting. Ungulates concentrate around mangers, building up high densities there while the surrounding landscapes are depopulated. Predators preying on those species are faced with a depletion of food in large parts of their ranges. Accordingly, even the predators concentrate their foraging around the ungulate feeding sites. The encounter rate between predator and prey is much higher at these sites than usual, and hunting success is increased accordingly. The impact of predators on the ungulates can be very pronounced at these sites. Gossow and Honsig-Erlenburg (1986, cited in Breitenmoser and Haller 1993) reported that lynx (*Lynx lynx*) in Austria preyed heavily upon red deer (*Cervus elaphus*) that were concentrated at feeding stations. Anecdotal reports of lynx preying on roe deer (*Capreolus capreolus*) (collected by M. Schneider) and white-tailed deer (*Odocoileus virginianus*) (I. Häkkinen, pers. comm.) at winter feeding sites are widespread in Sweden and Finland, and this behavior is also reported in Switzerland (Breitenmoser and Haller 1993; Haller and Breitenmoser 1986).

Haller (1992) and Breitenmoser and Haller (1993) described in detail the effects of reintroduction of the lynx on roe deer and chamois (*Rupicapra rupicapra*) populations in Switzerland. Ungulate population densities were very high due to current management practices, and prey were not behaviorally adapted to confront a large predator other than humans. As a result, the preequilibrium impact of lynxes was very severe, albeit temporarily.

The wintering habits of reindeer in Sweden create a similar situation. Reindeer owners concentrate herds of 1,000–2,000 animals in small areas, where feeding stations are provided in addition to the natural forage (M. Schneider, pers. obs.). Losses, especially to lynx, wolf, and wolverine (*Gulo gulo*) predation, can be severe (Bjärvall et al. 1990).

Heavy losses of ungulates inflicted by large predators often provoke conflicts with people. Owners and hunters of the prey animals usually demand control of the predator population (and/or financial compensation). At least in Europe, however, populations of the large predators nowadays are for the most part rather small and thus vulnerable to extinction (Bergström et al. 1993; Breitenmoser and Breitenmoser-Würsten 1990; Council of Europe 1989; Promberger and Schröder 1993). Therefore, these large carnivores are protected in many countries. In general, conflicts between protection needs and economic interests are balanced against the carnivores. The insights gained from landscape ecology may help us to devise a spatial array of predator and prey populations that would minimize economic losses while ensuring survival of a viable predator population.

Summary

In this chapter, we summarize the development of the hypothesis of exploitation ecosystems from the 1981 model, which was built on the tacit premise of infinitely large and homogeneous habitat patches with self-contained dynamics, to later models that take into account movements of consumers among habitat patches with different primary productivities. The 1981 model is robust if habitat selection is "ideal free" and there is no opportunistic foraging during transit stretches. With "ideal despotic" habitat selection or opportunistic foraging, the most productive habitat of the landscape tends to drive local dynamics even in less productive habitats. The strength of this "spillover predation" depends on the relative abundance of different habitat types. If the productive habitat type covers only a relatively small fraction of the landscape, then deviations from the predictions of the 1981 model will not be large. However, if the productive habitat type predominates, then differences in local predation pressure may not even be in the direction predicted by the 1981 model. Abundance relationships among different habitat types are today influenced largely by human activities. Unexpected secondary changes in habitats may be generated by variations in the strength of spillover predation. Conserving the spatial structure of the landscape may thus be an essential part of nature protection. These ideas are illustrated by results from a long-term study on predator-prey relations in northern Norway.

Acknowledgments

We are grateful to Villem Aruoja, Toomas Armulik, Lauri Oksanen, and Elisabeth Wiklund for their invaluable help in the tracking work, and to Per Ekerholm for discussing his unpublished data with us. Ü. Rammul assisted with the trapping. Birger Hörnfeldt, Bill Lidicker, and Lauri Oksanen made valuable comments on the manuscript. We especially thank Bill Lidicker for his editorial help.

Literature Cited

Ambuel, B., and S. A. Temple. 1983. Area-dependent changes in the bird communities and vegetation of southern Wisconsin forests. *Ecology* 64:1057–68.

Andrén, H. 1994. Effects of habitat fragmentation on birds and mammals in landscapes with different proportions of suitable habitat: A review. *Oikos* 71:355–66.

Andrén, H., and P. Angelstam. 1988. Elevated predation rates as an edge effect in habitat islands: Experimental evidence. *Ecology* 69:544–7.

Andrén, H., P. Angelstam, E. Lindström, and P. Widén. 1985. Differences in predation pressure in relation to habitat fragmentation: An experiment. *Oikos* 45:273–7.

Angelstam, P., E. Lindström, and P. Widén. 1984. Role of predation in short-term population fluctuations of some birds and mammals in Fennoscandia. *Oecologia* 62:199–208.

Angelstam, P., E. Lindström, and P. Widén. 1985. Synchronous short-term population fluctuations of some birds and mammals in Fennoscandia. *Holarctic Ecol.* 8:285–98.

Bergström, M.-R., B. Terje, R. Franzén, G. Henriksen, M. Nieminen, Ø. Overrein, and O. M. Stensli. 1993. *Björn, Gaupe, Jerv och Ulv på Nordkalotten.* Nordkalottkommitéens Rapportserie, Rapport 30. Rovaniemi, Finland: Nordkalottkommitée. 56 pp.

Bjärvall, A., R. Franzén, M. Nordkvist, and G. Åhman. 1990. *Renar och Rovdjur.* Solna, Sweden: Naturvårdsverket. 296 pp.

Boitani, L., and P. Ciucci. 1993. Wolves in Italy: Critical issues for their conservation. Pages 75–90 *in* Promberger, C., and W. Schröder (eds.), *Wolves in Europe—Status and Perspectives.* Munich: Munich Wildlife Society. 136 pp.

Breitenmoser, U., and C. Breitenmoser-Würsten. 1990. *Status, Conservation Needs and Reintroduction of the Lynx* (Lynx lynx) *in Europe.* Nature and Environment Series No. 45. Strasbourg: Council of Europe. 47 pp.

Breitenmoser, U., and H. Haller. 1993. Patterns of predation by reintroduced European lynx in the Swiss Alps. *J. Wildl. Manage.* 57:135–44.

Bright, P. W. 1993. Habitat fragmentation—Problems and predictions for British mammals. *Mammal Rev.* 23:101–11.

Brown, J. L., and M. L. Rosenzweig. 1986. Habitat selection in slowly regenerating environments. *J. Theor. Biol.* 123:151–71.

Charnov, E. L. 1976. Optimal foraging, the marginal value theorem. *Theor. Popul. Biol.* 9:129–36.

Comins, H. N., M. P. Hassell, and R. M. May. 1992. The spatial dynamics of host-parasitoid systems. *J. Anim. Ecol.* 61:735–48.

Council of Europe. 1989. *Workshop on the Situation and Protection of the Brown Bear* (Ursus arctos) *in Europe.* Environmental Encounters Series No. 6. Strasbourg: Council of Europe. 80 pp.

Crowley, P. H. 1981. Dispersal and the stability of predator-prey interactions. *Am. Nat.* 118:673–701.

Diekmann, O., J. A. J. Metz, and M. W. Sabelis. 1989. Reflections and calculations on a predator-prey-patch problem. *Acta Applicandae Mathematicae* 14:23–35.

Doncaster, C. P., and J. R. Krebs. 1993. The wider countryside—Principles underlying the responses of mammals to heterogeneous environments. *Mammal Rev.* 23:113–20.

Erlinge, S. 1977. Spacing strategy in stoat *Mustela erminea. Oikos* 28:32–42.

Erlinge, S., and M. Sandell. 1988. Coexistence of stoat, *Mustela erminea,* and weasel, *M. nivalis:* Social dominance, scent communication, and reciprocal distribution. *Oikos* 53:242–6.

Free, C. A., J. R. Beddington, and J. H. Lawton. 1977. On the inadequacy of simple models of mutual interference for parasitism and predation. *J. Anim. Ecol.* 46:543–54.

Fretwell, S. D. 1972. *Populations in a Seasonal Environment.* Princeton, N.J.: Princeton University Press. 217 pp.

Fretwell, S. D. 1977. The regulation of plant communities by food chains exploiting them. *Perspect. Biol. Med.* 20:169–85.

Fretwell, S. D. 1987. Food chain dynamics: The central theory of ecology? *Oikos* 50:291–301.

Fretwell, S. D., and H. L. Lucas Jr. 1970. On territorial behavior and other factors influencing habitat distribution in birds. I. Theoretical development. *Acta Biotheor.* 19:16–36.

Gaines, M. S., N. C. Stenseth, M. L. Johnson, R. A. Ims, and S. Bondrup-Nielsen. 1991. A response to solving the enigma of population cycles with a multifactorial perspective. *J. Mammal.* 72:627–31.

Gossow, H., and P. Honsig-Erlenburg. 1986. Management problems with re-introduced lynx in Austria. Pages 77–83 *in* Miller, S. D., and D. D. Everett (eds.), *Cats of the World: Biology, Conservation, and Management.* Washington, D.C.: National Wildlife Federation. 501 pp. (As cited in Breitenmoser and Haller 1993)

Gyllenberg, M., and I. Hanski. 1992. Single species metapopulation dynamics: A structural model. *Theor. Popul. Biol.* 42:35–61.

Hairston, N. G., Jr., and N. G. Hairston Sr. 1993. Cause-effect relationships in energy flow, trophic structure, and interspecific interactions. *Am. Nat.* 142:379–411.

Hairston, N. G., F. E. Smith, and L. B. Slobodkin. 1960. Community structure, population control and competition. *Am. Nat.* 94:421–5.

Haller, H. 1992. *Zur Ökologie des Luchses* Lynx lynx *im Verlauf seiner Wiederansiedelung in den Walliser Alpen.* Mammalia depicta 15. Hamburg: Verlag Paul Parey. 62 pp.

Haller, H., and U. Breitenmoser. 1986. Der Luchs in der Schweiz - 14 Jahre nach seiner Rückkehr. Pages 30–4 *in* W. d'Oleire-Oltmanns (ed.), *Das Bärenseminar.* Berchtesgaden, Germany: Nationalparkverwaltung. 62 pp.

Hanski, I. 1991. Single-species metapopulation dynamics: Concepts, models and observations. Pages 17–38 *in* Gilpin, M., and I. Hanski (eds.), *Metapopulation Dynamics.* London: Academic Press. 336 pp.

Hanski, I., P. Turchin, E. Korpimäki, and H. Henttonen. 1993. Population oscillations of boreal rodents: Regulation by mustelid predators leads to chaos. *Nature (London)* 364:232–5.

Hansson, L. 1977a. Landscape ecology and stability of populations. *Landscape Plann.* 4:85–93.

Hansson, L. 1977b. Spatial dynamics of field voles *Microtus agrestis* in heterogeneous landscapes. *Oikos* 29:539–44.

Hansson, L., and H. Henttonen. 1985. Gradients in density variations of small rodents: The importance of latitude and snow cover. *Oecologia* 67:394–402.

Hassell, M. P., and S. W. Pacala. 1990. Heterogeneity and the dynamics of host-parasitoid interactions. *Philos. Trans. R. Soc. London B* 330:203–20.

Hastings, A. 1977. Spatial heterogeneity and the stability of predator-prey systems. *Theor. Popul. Biol.* 14:380–95.

Henttonen, H., T. Oksanen, A. Jortikka, and V. Haukisalmi. 1987. How much do weasels shape microtine cycles in the northern Fennoscandian taiga? *Oikos* 50:353–65.

Hobbs, R. J. 1993a. Fragmented landscapes in western Australia: Introduction. *Biol. Conserv.* 64:183–4.

Hobbs, R. J. 1993b. Effects of landscape fragmentation on ecosystem processes in the Western Australian wheatbelt. *Biol. Conserv.* 64:193–202.

Holt, R. D. 1977. Predation, apparent competition, and the structure of prey communities. *Theor. Popul. Biol.* 28:181–208.

Holt, R. D. 1984. Spatial heterogeneity, indirect interactions and the coexistence of prey species. *Am. Nat.* 124:377–406.

Holt, R. D. 1985. Population dynamics in two patch environments: Some anomalous consequences of an optimal habitat distribution. *Theor. Popul. Biol.* 28:181–208.

Holt, R. D. 1987. Prey communities in patchy environments. *Oikos* 50:276–91.

Holt, R. D. 1993. Ecology at the mesoscale: The influence of regional processes on local communities. Pages 77–88 *in* Ricklefs, R. E., and D. Schluter (eds.), *Species Diversity in Ecological Communities.* Chicago: University of Chicago Press.

Jacobs, M. 1988. *The Tropical Rain Forest.* Berlin: Springer-Verlag. 295 pp.

Kareiva, P. 1990. Population dynamics in spatially complex environments: Theory and data. *Philos. Trans. R. Soc. London B* 330:175–90.

Kruuk, H. 1972. *The Spotted Hyena.* Chicago: University of Chicago Press. 335 pp.

Lidicker, W. Z., Jr. 1975. The role of dispersal in the demography of small mammals. Pages 103–28 *in* Golley, F. B., K. Petrusewicz, and L. Ryszkowski (eds.), *Small Mammals: Their Productivity and Population Dynamics.* Cambridge: Cambridge University Press. 451 pp.

Lidicker, W. Z., Jr. 1988. Solving the enigma of microtine "cycles." *J. Mammal.* 69:225–35.

Lindström, E. 1989. The role of medium-sized carnivores in the Nordic boreal forest. *Finn. Game Res.* 46:53–63.

Łomnicki, A. 1978. Individual differences between animals and the natural regulation of their numbers. *J. Anim. Ecol.* 47:461–75.

MacArthur, R. M., and R. Levins. 1964. Competition, habitat selection, and character displacement in a patchy environment. *Proc. Natl. Acad. Sci. USA* 51:1207–10.

Martinsson, B., L. Hansson, and P. Angelstam. 1993. Small mammal dynamics in adjacent landscapes with varying predator communities. *Ann. Zool. Fenn.* 30:31–42.

McNaughton, S. J. 1979. Grazing as an optimization process: Grass-ungulate relationships in the Serengeti. *Am. Nat.* 132:87–106.

Mech, L. D. 1966. *The Wolves of Isle Royale.* Fauna Series 7. Washington, D.C.: U.S. National Park Service. 210 pp.

Moen, J., P. Lundberg, and L. Oksanen. 1993. Lemming grazing on snowbed vegetation during a population peak, northern Norway. *Arctic Alpine Res.* 25:130–5.

Morris, D. W. 1987. Spatial scale and the cost of density-dependent habitat selection. *Evol. Ecol.* 1:379–88.

Morris, D. W. 1991. On the evolutionary stability of dispersal to sink habitats. *Am. Nat.* 137:907–11.

Morris, D. W. 1994. Habitat matching: Alternatives and implications to populations and communities. *Evol. Ecol.* 8:387–406.

Oksanen, L. 1980. Abundance relationships between competitive and grazing-tolerant plants in productivity gradients in Fennoscandian mountains. *Ann. Bot. Fenn.* 17:410–29.

Oksanen, L., S. D. Fretwell, J. Arruda, and P. Niemalä. 1981. Exploitation ecosystems in gradients of primary productivity. *Am. Nat.* 118:240–61.

Oksanen, L., and T. Oksanen. 1981. Lemmings (*Lemmus lemmus*) and grey sided voles (*Clethrionomys rufocanus*) in interaction with their resources and predators on Finnmarksvidda, northern Norway. *Rep. Kevo Subarct. Res. Stn.* 17:7–31.

Oksanen, L., and T. Oksanen. 1992. Long-term microtine dynamics in north Fennoscandian tundra: The vole cycle and the lemming chaos. *Ecography* 15:226–36.

Oksanen, L., T. Oksanen, P. Ekerholm, J. Moen, P. Lundberg, L. Bondestad, M. Schneider, V. Aruoja, and T. Armulik. 1995. Structure and dynamics of arctic-subarctic grazing webs in relation to primary productivity. *In* Winemiller, K., and G. Polis (eds.), *Food Webs: Integrating Patterns and Processes.* London: Chapman & Hall. In press.

Oksanen, T. 1990a. Exploitation ecosystems in heterogeneous habitat complexes. *Evol. Ecol.* 4:220–34.

Oksanen, T. 1990b. *Predator-Prey Dynamics in Small Mammals along Gradients of Primary Productivity.* Ph.D. thesis. Umeå, Sweden: University of Umeå. 184 pp.

Oksanen, T. 1993. Does predation prevent Norwegian lemmings from establishing permanent populations in lowland forests? Pages 425–37 *in* Stenseth, N. C., and R. A. Ims (eds.), *The Biology of Lemmings.* Linnean Soc. Symp. Ser. No. 15. London: Academic Press. 683 pp.

Oksanen, T., L. Oksanen, and M. Gyllenberg. 1992a. Exploitation ecosystems in heterogeneous habitat complexes. II: Impact of small-scale heterogeneity on predator-prey dynamics. *Evol. Ecol.* 6:383–98.

Oksanen, T., L. Oksanen, and M. Norberg. 1992b. Habitat use of small mustelids in north Fennoscandian tundra: A test of the hypothesis of patchy exploitation ecosystems. *Ecography* 15:237–44.

Oksanen, T., M. Power, and L. Oksanen. 1995. Ideal free habitat selection and consumer resource dynamics. *Am. Nat.* In press.

Ostfeld, R. S. 1992. Small-mammal herbivores in a patchy environment: Individual strategies and population responses. Pages 43–74 *in* Hunter, M. D., T. Ohgushi, and P. W. Price (eds.), *Effects of Resource Distribution on Animal-Plant Interactions.* New York: Academic Press. 505 pp.

Promberger, C., and W. Schröder (eds.). 1993. *Wolves in Europe—Status and Perspectives.* Munich: Munich Wildlife Society. 136 pp.

Pulliainen, E. 1981. Winter habitat selection, home range, and movements of the pine marten (*Martes martes*) in a Finnish Lapland forest. Pages 1068–87 *in* Chapman, J. A., and D. Pursley (eds.), *Worldwide Furbearer Conference Proceedings,* vol. 2. Frostburg, Md. 2,056 pp.

Pulliam, H. R. 1988. Sources, sinks, and population regulation. *Am. Nat.* 132:652–61.

Pulliam, H. R., and B. J. Danielson. 1991. Sources, sinks, and habitat selection: A landscape perspective on population dynamics. *Am. Nat.* (*Suppl.*) 137:S50–66.

Reeve, J. D. 1988. Environmental variability, migration, and persistence in host-parasitoid systems. *Am. Nat.* 132:810–36.

Reeve, J. 1990. Stability, variability, and persistence in host-parasitoid systems. *Ecology* 71:422–6.

Rosenzweig, M. L. 1971. Paradox of enrichment: Destabilization of exploitation ecosystems in ecological time. *Science* 171:385–7.

Rosenzweig, M. L. 1973. Exploitation in three trophic levels. *Am. Nat.* 107:275–94.

Rosenzweig, M. L., and R. H. MacArthur. 1963. Graphical representation and stability conditions of predator-prey interactions. *Am. Nat.* 97:209–33.

Sabelis, M. W., and O. Diekmann. 1988. Overall population stability despite local extinction:

The stabilizing influence of prey dispersal from predator-invaded patches. *Theor. Popul. Biol.* 34:169–76.

Sabelis, M. W., O. Diekmann, and V. A. A. Jansen. 1991. Metapopulation persistence despite local extinction: Predator-prey patch models of the Lotka-Volterra type. Pages 267–83 *in* Gilpin, M., and I. Hanski (eds.), *Metapopulation Dynamics.* London: Academic Press. 336 pp.

Saunders, D. A., R. J. Hobbs, and G. W. Arnold. 1993. The Kellerberrin Project on fragmented landscapes: A review of current information. *Biol. Conserv.* 64:185–92.

Schaller, G. B. 1972. *The Serengeti Lion.* Chicago: University of Chicago Press. 480 pp.

Sinclair, A. R. E. 1977. *The African Buffalo: A Study of Resource Limitation of Populations.* Chicago: University of Chicago Press. 355 pp.

Sonerud, G. A. 1986. Effect of snow cover on seasonal changes in diet, habitat, and regional distribution of raptors that prey on small mammals in boreal zones of Fennoscandia. *Holarctic Ecol.* 9:33–47.

Stephens, D. W., and J. R. Krebs. 1986. *Foraging Theory.* Princeton, N.J.: Princeton University Press. 247 pp.

Taylor, A. D. 1988. Large-scale spatial structure and population dynamics in arthropod predator-prey systems. *Ann. Zool. Fenn.* 25:63–74.

Taylor, A. D. 1990. Metapopulations, dispersal, and predator-prey dynamics. *Ecology* 71:429–33.

Wielgolaski, F. E. 1975. Primary production of alpine meadow communities. Pages 121–8 *in* Wielgolaski, F. E. (ed.), *Fennoscandian Tundra Ecosystems,* Part 1. Ecological Studies 16. Berlin: Springer-Verlag. 366 pp.

Wiens, J. A., J. F. Addicott, T. J. Case, and J. Diamond. 1986. Overview: The importance of spatial and temporal scale in ecological investigations. Pages 145–53 *in* Diamond, J., and T. J. Case (eds.), *Community Ecology.* New York: Harper & Row. 665 pp.

Zimen, E. 1990a. *Wildwege Europas.* Munich: Knesebeck und Schuler. 134 pp.

Zimen, E. 1990b. *Der Wolf. Verhalten, Ökologie und Mythos.* Munich: Knesebeck und Schuler. 456 pp.

Zonneveld, I. S., and R. T. T. Forman (eds.). 1990. *Changing Landscapes: An Ecological Perspective.* New York: Springer-Verlag. 286 pp.

Part III

Model Systems: An Experimental Protocol

We think of landscapes as generally pretty substantial chunks of the biosphere. Moreover, every place on Earth is inherently unique. How then is it possible to do controlled experiments in landscape ecology? In Part III, two chapters explore this dilemma, and both conclude that experimental approaches at the landscape level are not only feasible, they are also critically important to our understanding of mammalian ecology and conservation.

Researchers conducting controlled experiments to elucidate landscape processes confront three major obstacles: spatial scale, replication, and predictability. If landscapes are viewed as ecological systems containing two or more community-types (Chap. 1), then the spatial scale required may be much less than that needed for studying landscapes on the level of human activities. The relevant spatial scale is a function of the organisms of primary focus. For species of small mammals, meaningful investigations are possible at what has been termed the *mesocosm* scale.

Replication of treatments is a necessary part of any experimental protocol. The challenge for landscape-level experiments is not fundamentally different from that for experiments at any other level. No two experimental entities can ever be exactly alike, even nonliving ones; however, it is certainly often possible to make such entities very similar. The larger and more complex the entities, the harder it is to make them extremely similar. Therefore, as one goes up the hierarchy of biological organization and increases the spatiotemporal scale of the investigation, satisfactory replication becomes increasingly difficult to achieve. It is certainly not impossible, however, as the following chapters make abundantly clear.

Tied to the difficulty of replication is the question of predictability, a desirable goal of any experimental investigation. Predictability is a function of our understanding of the system under investigation and of its inherent regularity. Poor understanding or inherent irregularity of performance can both foil predictions, but either is sufficient. Landscapes are more complex than the communities that compose them, and so understanding them is correspondingly more difficult. Moreover, landscapes appear to be dominated by idio-

Experimental enclosures used to study landscape processes as they affect *Microtus oeconomus* at Evenstad, Østerdalen, Hedmark, Norway. (Photo by W. Z. Lidicker Jr., taken 3 August 1990.)

syncratic behaviors, which tend to obscure the generalities we seek in order to make predictions about future behavior or unstudied landscapes. This characteristic results not only from inherent complexity, but also because landscapes are relatively few in number. There are necessarily fewer landscapes than communities, and fewer communities than populations. Large sample sizes are thus difficult to obtain, especially of experimental replicates.

Given all these difficulties, even the avid experimentalist may well revert to organismal or lower levels of investigation. In fact, only a handful of investigators have had the courage, and the resources, to tackle experimental landscape ecology. However, real power lies in the experimental approach, and the chapters in this part exemplify the best of these efforts so far.

One additional major program of this type is that led by N. C. Stenseth and R. A. Ims of the University of Oslo. They were part of our original symposium but unfortunately were unable to contribute a chapter to this volume. Because the community of such investigators is tiny, it is appropriate to mention two others who are doing similar research: R. S. Ostfeld at the Institute of Ecosystem Studies in Millbrook, New York, and M. A. Bowers at the Blandy Experimental Farm near Boyce, Virginia.

Chapter 8 summarizes the pioneering effort of Gary Barrett and his associates at Miami University in Oxford, Ohio. Using several species of small mammals as their focal organisms, they have successfully manipulated various landscape features, such as patch geometry, quality, resource dispersion, corridors, and interfragment distances. The discerning reader will notice that in this chapter, the term *ecosystem* is used, as is commonly done, to define a level of biological complexity between those of the community and the landscape. One of the main thrusts of this book is to explore the alternative view (Chap. 1) that the term *landscape (seutu)* can be appropriately used for the next highest level of complexity above that of communities. According to this perspective, landscapes are defined so as to encompass new qualities of biological complexity and hence new emergent properties; the term ecosystem has never been defined in this way but is instead a gambit for encouraging thinking in the context of systems. Constructive discussion of this important issue requires the expression of various viewpoints.

In the second chapter in this series (Chap. 9), four authors collaborate in synthesizing the results of two long-term studies of the small-mammal community living in the grasslands of eastern Kansas. The results from one study, which involved a large block of prairie vegetation, are compared and contrasted with those of the other study, which used contrived habitat patches of various sizes. The authors document the important insight that the effects of fragmentation are different for each species studied and may even vary among sex and age cohorts in the same species. These authors are enthusiastic about the value of experimental manipulations in elucidating landscape processes.

Finally, I offer a few comments on the experimental approach taken by Stenseth and Ims mentioned above. Their complex four-year investigation (1990–93), with *Microtus oeconomus* as the model organism, also supports the efficacy of studying small mammals in experimentally manipulated landscapes. Not only was the program part of our original symposium, but I was privileged to participate in this research, especially in 1990 and 1992. The experiments utilized seven 0.5 ha enclosures located at Evenstad in Østerdalen, Hedmark, Norway (see the accompanying photos). Within the enclosures, the lush river bottom grasslands, made homogeneous by sowing of three species of grasses plus *Trifolium,* were mowed to create habitat patches. Vole responses to diminishing patch size were studied as well as to manipulation of patch size and presence or absence of corridors. One enclosure included a gradient of patch sizes. A particularly intriguing feature of the program was

View of Østerdalen at Evenstad, Hedmark, Norway, showing
natural landscape features (riparian zone, altitudinal zonation
of vegetation), anthropogenic modifications (timber harvesting, agriculture), and experimental enclosures. (Photo by W.
Z. Lidicker Jr., taken 12 June 1990.)

the use of two subspecies of *M. oeconomus* that differ in characteristics of
their social behavior as well as in morphology. Treatments were replicated
within years and duplicated in a second year. Stenseth and Ims anticipate that
the study of model systems such as this will be useful in extrapolating to
larger-scale landscapes that are less amenable to experimental manipulation.

8

Reflections on the Use of Experimental Landscapes in Mammalian Ecology

Gary W. Barrett, John D. Peles, and Steven J. Harper

Landscape Ecology

Landscape ecology has emerged during the past decade as a paradigm within the realm of applied ecology (Barrett 1992; Barrett and Bohlen 1991). This integrative paradigm is concerned with (a) the development and dynamics of landscape patterns, (b) the exchanges and spatiotemporal interactions of biotic and abiotic materials across the landscape mosaic, (c) the manner in which biotic and abiotic processes are influenced by spatial heterogeneity (i.e., landscape pattern), and (d) the management of spatial heterogeneity for the long-term benefit and survival of society (Risser et al. 1984).

The study of landscape ecology requires an understanding of hierarchy theory, which allows the ordering of discrete elements into increasingly inclusive and complex sets (Allen and Starr 1982; Urban et al. 1987). The levels-of-organization concept (Allen and Hoekstra 1992) is an excellent model for understanding hierarchy theory. Within the framework of this concept, the landscape level of organization includes patterns and processes that are observable at lower organizational levels (i.e., cellular through ecosystem levels) in addition to phenomena unique to the landscape level. Thus, a landscape perspective allows integration across all levels of organization for the study of patterns and processes at varying spatial and temporal scales in a hierarchical manner. We argue that mammalogists increasingly need to address questions and conduct experimental studies involving mammals across these levels of hierarchy, especially at the landscape level.

Mammalogists have recently recognized the importance of conducting research at the landscape level of organization (Lidicker et al. 1992). Whereas few studies have focused on large mammals (Bissonette and Broekhuizen,

Chap. 6; Diamond 1993), small mammals have recently been the subject of a number of landscape-level studies. These studies may be grouped into two basic research approaches. Beginning with Hansson (1977) and Wegner and Merriam (1979), several small-mammal landscape-level investigations have focused on the effects of naturally occurring landscape patterns on mammalian population dynamics (Fahrig and Merriam 1985; Henderson et al. 1985; Kozakiewicz 1993; Laurance, Chap. 3; Lidicker et al. 1992; Merriam, Chap. 4; Oksanen and Schneider, Chap. 7; Szacki et al. 1993). In the second approach, a replicated research design is used to manipulate landscape elements (e.g., corridors) as variables within an experimental landscape design (Diffendorfer et al., Chap. 9; Harper et al. 1993; La Polla and Barrett 1993; Lorenz and Barrett 1990; Robinson et al. 1992; N. C. Stenseth and R. A. Ims, pers. comm.).

In this chapter, we discuss the experimental landscape approach to mammalian ecology and give examples of simulated landscape studies in which landscape variables have been manipulated. We also discuss applications of this approach and the need for future experimental studies in mammalian ecology that couple theory with practice.

Experimental Landscape Approaches

Few question the validity and rigor of a hypothesis-testing approach to fields of science such as mammalian ecology. Although there is an acknowledged need for additional experimental approaches to analyzing natural systems (e.g., cybernetics, problem-solving algorithms, and cost-benefit analysis [Barrett 1985, 1993]), the scientific method remains, perhaps, the most rigorous approach to the investigation of natural phenomena, including the role of mammals functioning within natural habitats of varying scales. This approach requires the formulation of a succinct null hypothesis and a research design that includes replication of the experimental variable being manipulated (e.g., stressor, habitat element, or resource) as well as replication of the level of organization and/or habitat involved.

Recently, much emphasis has been placed on the importance of investigations that transcend and attempt to integrate science at several levels of organization based on a hierarchical perspective (O'Neill et al. 1986). Such studies also frequently address questions at greater temporal or spatial scales. Urban et al. (1987) called attention to the need for a hierarchical perspective in the area of landscape ecology, especially with regard to the role of spatial patterns and their effect on natural phenomena.

Although numerous replicated, experimental studies in mammalian ecology have focused on the population, community, and ecosystem levels (Barrett 1988; Hall et al. 1991), few experimental studies have addressed questions in mammalian ecology at the landscape level (but see Harper et al. 1993). To date, most landscape-level studies have lacked a robust research design, frequently disregarding systems replication and thereby making it impossible to test for differences among treatments.

Hurlbert (1984) stressed the importance of replication at the ecosystem level. He chastised the "pseudoreplication" of ecosystem-level studies (Crowner and Barrett 1979) but praised nonreplicated, manipulated landscape (watershed) level investigations at the Hubbard Brook deforestation experiment (Likens et al. 1970, 1977). One could argue that it is too costly, or indeed impossible, to replicate landscape-level phenomena. We argue here, however, that it is economically feasible and ecologically important to address landscape-level questions by manipulating landscape elements within replicated, simulated landscape designs.

We will describe experimental approaches to the study of mammalian ecology at the landscape level, in which we attempt to simulate and/or replicate landscape habitats, especially the landscape matrix, while manipulating landscape elements (e.g., corridors), patch area (e.g., habitat fragmentation), patch geometry (e.g., edge-area ratios), or patch quality (e.g., patch cover or food quality). This strategy may also be viewed as an experimental mesocosm approach to natural (macrocosm) landscapes. It should yield new insights into aspects of mammalian ecology, including, for example, the role of corridors in patch connectivity, dispersal behavior, and/or genetic drift between and within patches; and the importance of patch quality, patch geometry, and habitat fragmentation in mammalian population dynamics.

Although significant human and financial resources may be needed initially to establish and maintain such complex investigations (e.g., experimental enclosures, farm equipment, or large plots of land), we argue that these approaches provide information that is vital to increasing our understanding of how mammalian populations function in natural landscapes. The simulated landscape approach not only addresses questions and tests hypotheses at the landscape level, but also simultaneously provides opportunities for studies focused on population, community, and ecosystem levels. For example, we urgently need to learn more about the transfer of genetic information and/or abiotic resources between ecosystems (Risser et al. 1984), how patch quality (or architecture) influences plant-animal community-level coevolution (Stueck and Barrett 1978), and how patch quality affects dispersal behavior (Harper et al. 1993).

We believe that the simulated landscape-level approach also provides insight into questions at greater temporal and spatial scales—realistic scales that mammals function within daily and have evolved within through eons of time. It also acknowledges the role of human intervention in phenomena such as habitat fragmentation, changes in landscape configurations, ecological energetics and resource cycling, and explores as well how corridors function simultaneously as barriers to and conduits for the transfer of genetic materials and/or resources. A landscape approach does not detract from the importance of studies that focus on lower levels of organization, but rather it promotes a broader understanding of mammalian ecology at the "real world" landscape level. Although we will never be able to replicate the global scale, we can (and should) design and conduct studies at the next hierarchical level of organization above the ecosystem.

Simulated Landscapes: An Experimental Approach

Landscape as a Habitat Mosaic

Ecological processes occurring within natural landscapes are often complex; they reflect the effects of multiple interacting factors operating at different spatial and temporal scales. For example, complexity arises as a result of the dynamic nature of habitat patches, which is a result of interactions among physical, biological, and social factors (Turner 1989). Such complexity makes hypothesis testing and experimentation difficult.

To overcome ecological complexities associated with natural landscapes, we have adopted a different type of research design (La Polla and Barrett 1993; Lorenz and Barrett 1990). We conduct replicated studies in which landscape elements (e.g., patch quality, size, shape, connectivity, or degree of fragmentation) are manipulated as variables within a simulated homogeneous landscape mosaic. Thus, we can discern the direct effects of landscape structural elements on ecological processes because potentially confounding effects of other factors have been minimized or eliminated. Investigations of individual landscape elements set the framework for future studies by identifying the elements that are most likely to have direct effects on ecological processes.

Independent studies incorporating this experimental approach to date have investigated the effects of patch geometry, patch architecture, landscape corridors, and degree of fragmentation on population dynamics and dispersal movements of small mammals (Fig. 8.1). Sheet metal enclosures and habitat patches surrounded by mowed and/or tilled borders have been used to create experimental landscapes. These investigations have been conducted at the Miami University Ecology Research Center (ERC) near Oxford in Butler County, Ohio. The ERC, a nationally recognized center for basic and applied ecological research (Barrett 1987), encompasses 70 ha (170 acres) of diverse habitat types and has on-site laboratory facilities to enhance terrestrial research in integrative fields of study such as landscape ecology.

Habitat Patch Geometry

The geometry of habitat patches may affect population dynamics because patches of different sizes and shapes have different edge-to-area ratios. Several small patches have more habitat edge per unit of area than a single large patch of equal area, and elongated patches have more than round patches of equal area. Recent studies of plant and nongame animal species have revealed many characteristics of edges that are considered undesirable, including increased predation, parasitism, and herbivory (Alverson et al. 1988; Harris 1988; Laudenslayer 1986; Laurance, Chap. 3; Temple and Cary 1988; Yahner 1988). These effects will be manifested most strongly in patches with high edge-to-area ratios.

The geometry of uniformly high-quality habitat patches may also directly influence animal population dynamics and dispersal behavior. For example, Stamps et al. (1987a) demonstrated with simulation models that habitat geometry and edge permeability (i.e., edge "hardness") can have major effects on rates of emigration from uniformly high-quality patches. These models indicated that the proportion of home ranges at the edge of a patch is positively related to the proportion of dispersers emigrating from that patch. Furthermore, the model suggested that animals in small, elongated, "soft-edged" patches (i.e., patches bounded by habitat that emigrants are able to cross) are more likely to defend larger territories (Stamps and Buechner 1985) and to suffer fewer visits from intruders (Stamps et al. 1987b) than are animals in larger, square patches.

To determine the effects of patch shape on the dispersal behavior and population dynamics of the meadow vole (*Microtus pennsylvanicus*), Harper et al. (1993) conducted a field experiment replicated over two years employing four square (40 m × 40 m) and four rectangular (16 m × 100 m) habitat patches of equal area (1,600 m^2 [see Fig. 8.1a]). Predictions generated from three hypotheses were tested. First, the hypothesis that patches with high edge-to-area ratios cause increased dispersal rates (Stamps et al. 1987a) led us to predict that a greater proportion of individuals would disperse from rectangular patches than from square patches. Second, the hypothesis that

a)

Fig. 8.1. Simulated experimental landscapes that incorporate a replicated research design for examining the role of landscape elements in mammalian ecology: (a) landscape patch geometry (reprinted, by permission, from Harper et al. 1993); (b) patch architecture (J. D. Peles and G. W. Barrett, unpub. data); (c) patch connectivity (reprinted, by permission, from La Polla and Barrett 1993); (d) habitat fragmentation (R. J. Collins and G. W. Barrett, unpub. data).

b)

c)

d)

home ranges are larger within patches with high edge-to-area ratios (Stamps et al. 1987b) led us to predict that home ranges would be larger for residents in rectangular patches than for those in square patches. Third, because of the expected increases in dispersal rates and home range sizes, we hypothesized that fewer individuals could be supported within patches with high edge-to-area ratios; therefore, we predicted that rectangular patches would have lower population densities than square patches.

The results indicated that the numbers of dispersers, but not dispersal rates, were greater in rectangular patches than in square patches only when vole densities were low. Interestingly, home ranges were of equal size but different shape in the contrasting patch shapes. Population density, recruitment, disperser body mass, resident body mass, survival, and age structure were not affected by differences in patch shape. We concluded that patch shape does not markedly affect the population dynamics of *M. pennsylvanicus* at this spatial scale; rather, meadow voles appear to be an edge-tolerant species that is capable of changing the shape of its home range (Harper et al. 1993).

Landscape Patch Architecture

Manipulating the architecture of landscape patches or habitat resources is another example of an experimental landscape approach to investigating the role of landscape elements (or resources contained therein) in mammalian ecology. We define patch architecture as the spatial and temporal arrangement of abiotic (e.g., moisture or light) and biotic (e.g., food or vegetative cover) habitat components. Thus, this term may refer to overall patch quality, the homogeneity of resources in a single patch, or short-term (growing season) and long-term (successional) changes in landscape patches.

Effects of resource partitioning on small-mammal population dynamics have been examined at the experimental ecosystem level. For example, Stueck and Barrett (1978) varied the distribution of food resources within simulated grassland experimental ecosystems to examine the effects of food (resource) partitioning on the population dynamics of the house mouse (*Mus musculus*). Four of the grids contained a centralized food (corn) depot, while the other four contained equally spaced, decentralized depots. Populations in centralized grids reached a peak density of about 20 animals per grid, compared to about 30 animals per grid in the decentralized grids. In addition, house mice of both sexes had significantly greater survival in patches with decentralized food resources relative to those with centralized food resources.

Maly et al. (1985) used a similar design to investigate the effects of resource partitioning on dispersal behavior of house mice. Significantly more house mice dispersed from centralized grids than from decentralized grids. This behavior was attributed to more rigid social hierarchies in centralized grids. Thus, these results indicate that the study of small-mammal populations in experimental landscape mosaics must consider not only the quantity and quality of resources within patches, but also the spatial arrangement of these resources.

A recent experimental design involving landscape patch architecture has permitted us to investigate the effects of habitat quality on the population dynamics and dispersal behavior of *M. pennsylvanicus* and to compare seasonal and yearly changes in patch quality (J. D. Peles and G. W. Barrett, unpub. data). Twelve experimental grassland patches (Fig. 8.1b), consisting of a 20 m × 20 m interior of grassland vegetation surrounded by a 1.5 m-wide mowed perimeter, were established. Each experimental patch was surrounded by a drift fence 23 cm high. Vegetative cover was manipulated to vary habitat quality during this two-year investigation. Ground litter and aboveground biomass were periodically removed to maintain the four reduced-cover treatments. Ground litter plus 1.25 m × 1.25 m plywood squares were added to each of the four enhanced-cover treatment patches. The four control treatments were not manipulated.

Our findings illustrate the necessity for considering seasonal and yearly changes in patch quality and patch architecture. For example, although no differences were observed in population dynamics or dispersal behavior of voles during the peak in the growing season (June-August), vegetative cover was found to be extremely important during early spring and late fall. In addition, differences in patch quality between years were responsible for differences in population dynamics and dispersal behavior among treatments (J. D. Peles and G. W. Barrett, unpub. data).

Landscape Connectivity

Computer models and natural corridors have been used to investigate the importance of landscape corridors for populations of mammals (Merriam, Chap. 4). The presence of corridors between habitat patches has been hypothesized to enhance population growth rates for species that use habitat corridors as dispersal routes (Fahrig and Merriam 1985). Wegner and Merriam (1979) and Henderson et al. (1985) found that white-footed mice (*Peromyscus leucopus*) and eastern chipmunks (*Tamias striatus*) experience frequent local extinctions in relatively isolated patches of habitat and depend for survival on fluxes of colonists using habitat corridors among patches.

Corridors also vary in quality, as determined by differences in survival rates of dispersers within corridors (Merriam and Lanoue 1990). Differential survivorship during dispersal movements can affect population dynamics within the mosaic of patches through changes in immigration rates. Experimental corridor manipulations have quantified the effects of corridor type (Lorenz and Barrett 1990) and corridor width (La Polla and Barrett 1993) on small-mammal population dynamics.

Lorenz and Barrett (1990), for example, investigated the use of contrasting types of corridors (human-made vs. natural) by house mice (*Mus musculus*). Strips of old-field vegetation, either alone or in combination with split-rail fencing, were established to simulate natural and human-made corridors, respectively. Dispersal movements of house mice were monitored in replicated trials conducted in simulated agricultural landscapes during each season of the year. Significantly more house mice used the corridors with split-rail fencing to disperse during late summer and autumn trials. Both types of corridors influenced the dispersal behavior of *Mus,* since no house mice were captured in the matrix.

La Polla and Barrett (1993) used mowing and tilling to create experimental landscape patches in order to examine the effects of corridor presence and corridor width on the dispersal behavior and population dynamics of the meadow vole (*M. pennsylvanicus*). Each replicate contained six 20 m × 20 m habitat patches arranged into blocks of three treatments each (Fig. 8.1c). Pairs of patches were either unconnected to one another, were connected by a 1 m-wide natural vegetation corridor, or were connected by a 5 m-wide natural vegetation corridor. Tracking tubes and live traps were used to monitor dispersal movements. Because voles have been demonstrated to be reluctant to cross unfavorable habitat strips (Cole 1978; Swihart and Slade 1984), habitat patches connected with corridors were expected to sustain higher rates of dispersal than unconnected patches. In addition, wide corridors were hypothesized to be of higher quality than narrow corridors; therefore, more dispersal movements were predicted between patches connected by 5 m corridors than between those connected by 1 m corridors.

Significantly more male voles dispersed between patches that were connected by corridors than between unconnected patches; however, rates of dispersal between patches connected by wide and narrow corridors were similar. Although tracking tube data indicated that most movements were confined to corridors (i.e., alpha dispersal), some movement was detected between adjacent treatments within a replicate block (beta dispersal) and between blocks (gamma dispersal). These findings indicated that dispersal behavior of meadow voles was affected by, but not restricted to, landscape corridors. Interestingly, vole dispersal was not related to corridor width at this spatial scale.

Habitat Fragmentation

Habitat fragmentation involves a reduction in habitat area as well as the isolation of patch elements within the total landscape (Wilcove et al. 1985). Therefore, patch size and spacing can be manipulated experimentally to study the role of patchiness in mammalian ecology. For example, a recent investigation at the ERC addressed the question of how habitat fragmentation affects the population dynamics of *M. pennsylvanicus* (R. J. Collins and G. W. Barrett, unpub. data). The research design involved the creation of simulated landscape patches within eight 0.1 ha enclosures. Four of the enclosures contained a 160 m^2 patch to represent nonfragmented habitat, and the other four each contained four 40 m^2 patches to represent fragmented habitat (Fig. 8.1d).

Vole population densities differed between treatments during October, when fragmented patches contained greater vole densities than nonfragmented patches. Importantly, significantly more females were found within the fragmented treatment, which suggests that habitat fragmentation could affect reproductive success in natural grassland landscapes. No differences, however, were observed between treatments in rates of recruitment at this scale. Radio-telemetry data indicated that females in the nonfragmented treatment were more active in the matrix (i.e., poor-habitat area) than females in the fragmented treatment. This differential use of space by female voles between treatments suggests that habitat fragmentation may affect home range size in meadow voles.

Discussion

A Mesocosm Perspective in Mammalian Ecology

The research designs outlined above involve the manipulation of landscape elements or resources within simulated landscapes, and they embody a viable mesocosm approach that can be applied to future studies in mammalian ecology. This type of experimental approach can be a valuable research tool because it provides ease of replication on a manageable scale (Odum 1984). Mesocosms have been successfully used in several ecosystem-level studies in mammalian ecology (Barrett 1968, 1988; Diffendorfer et al., Chap. 9; N. C. Stenseth and R. A. Ims, pers. comm.). We propose that this perspective now be applied to large spatial scales.

Mesocosms represent a means by which reductionist and holistic approaches may be incorporated into integrative research programs (Odum 1984). The integration of these two approaches may provide new insights into how seasonal (i.e., short-term) and successional (long-term) landscape processes affect the ecology of small mammals. For example, ecosystem- and landscape-level processes may be investigated over long time periods, whereas shorter-term phenomena, such as the effects of landscape elements on vertebrate population dynamics (e.g., Robinson et al. 1992), may be studied in the time frame of graduate or undergraduate research studies.

Where Do We Go from Here?

Lubchenco et al. (1991) outlined a research strategy that focuses on the necessity of addressing questions and collecting information at the landscape and global scales, especially if we are to underpin sustainable development with sound ecological theory into the 21st century. Barrett (1992) stressed the need to integrate information occurring at greater temporal and spatial scales. Mammalian studies (excluding those mainly focused on *Homo sapiens*) have traditionally been designed to address questions at the population, community, and ecosystem levels, with less regard for questions pertaining to landscape dynamics. Because mammalian studies have frequently lacked proper replication, especially at the landscape level (Hurlbert, 1984; but see Barrett 1988; Hall et al. 1991; Robinson et al. 1992 for examples of replicated research designs at the simulated landscape level), we emphasize the need for proper replication at the landscape scale as well.

The Long-Term Ecological Research (LTER) sites sponsored by the National Science Foundation (NSF) have promoted a watershed approach to integrate across levels of organization and temporal/spatial scales; however, these studies frequently minimize the human component, especially the ways in which human intervention has affected the landscape mosaic. Barrett (1992) suggested, for example, that the NSF LTER program design an agrolandscape-level study site that would encompass human communities (i.e., towns) within the landscape to provide a "real world" investigative perspective. This approach holds promise in central North America, which has a landscape dominated by agriculture in which mammals, for example, must function within increasingly smaller landscape patches. The effects of agricultural perturbations on mammalian ecology have been poorly investigated, partly because of the scarcity of human and financial resources required to undertake studies at this scale. Focus on traditional disciplines at the undergraduate and graduate levels of higher education has also impeded this integrative approach (Barrett 1993). Landscape approaches to research (natural and simulated) will likely act as a catalyst for these needed future investigations.

Further, there is an intellectual need first to combine theoretical (basic) and practical (applied) research activities; second, to advance from disciplinary or mainly taxon-based approaches to interdisciplinary or transdisciplinary research agendas; and third, to design long-term research studies that address questions at the landscape level. Hurlbert (1984) noted that studies in mammalian ecology were particularly guilty of designs that fail to provide necessary replication (pseudoreplication) at the ecosystem level; we do not want to repeat these mistakes at the landscape level.

Covey (1989) noted that society, including the science of ecology, must advance from dependence (intradisciplinary) to independence (disciplinary) to interdependence (transdisciplinary) stages of personal, educational, scientific, and societal maturity. We present evidence that landscape-level investigations illustrate this maturation process, in addition to helping a cadre of people (undergraduate and graduate students, postdoctoral scholars, senior investigators, and the public) work together as a team in designing and conducting these studies.

One goal, for example, will be to implement sustainable agriculture at the agrolandscape level (Barrett 1992). This implementation process will include an array of alternative farming practices within the agricultural landscape mosaic. Studies already have been conducted to clarify the role of natural and planted corridors (i.e., strip-cropping) in agricultural systems. Early results have focused mainly on insect dispersal (Kemp and Barrett 1989; Pavuk and Barrett 1993). Studies are now being conducted, however, to investigate how populations of mammals are affected by these alternative cultivation practices (Williams et al. 1994). These studies demonstrate the potential benefits from a simulated, replicated research design.

Our intention here is to encourage a broad array of landscape studies that can then be used to compare the response of mammals to landscape patterns and changes across taxonomic groups and niche guilds, to compare findings across ecosystem and/or biome boundaries, and to predict how human interventions will affect mammalian regulatory mechanisms. Such studies will not only deepen our theoretical base of understanding, but will also provide new information needed to help us manage our natural resources in a sustainable manner. These studies will thus help to bring us one step closer to realizing the objectives outlined in the Sustainable Biosphere Initiative (Lubchenco et al. 1991) sponsored by the Ecological Society of America.

Summary

Mammalian ecologists increasingly need to address questions at the landscape level of organization. Previous studies have frequently lacked a robust research design that includes experimental manipulations and/or systems replication. We outline examples of replicated research designs in which landscape elements, resources, patterns, or configurations are manipulated as variables within a simulated landscape mosaic. These designs have been used at the Miami University Ecology Research Center during the past decade, and the results have contributed to increased understanding of how small mammals function at the landscape level. This mesocosm approach to natural landscapes has been used to examine the importance of landscape elements (such as corridors or patches) on small-mammal population dynamics and dispersal behavior via the manipulation of variables such as patch geometry, habitat architecture, patch connectivity, and habitat fragmentation. Specifically, these studies have addressed questions regarding the role of patch geometry; seasonal and yearly differences in patch quality; resource allocation patterns within simulated patches; corridor type, width, and presence; and habitat fragmentation in the population dynamics and dispersal behavior of small mammals.

Although designing and maintaining experimental landscapes requires significant human and financial investment, this approach is essential to bettering our understanding of mammalian ecology at the landscape level. We suggest that this approach be implemented as part of long-term ecological research programs (such as LTER and the Environmental Protection Agency's Environmental Monitoring and Assessment Program [EMAP]) that promote research and monitoring at varying temporal and spatial scales, including the impact of humankind on landscape dynamics.

Literature Cited

Allen, T. F. H., and T. W. Hoekstra. 1992. *Toward a Unified Ecology.* New York: Columbia University Press. 384 pp.

Allen, T. F. H., and T. B. Starr. 1982. *Hierarchy: Perspectives for Ecological Complexity.* Chicago: University of Chicago Press. 310 pp.

Alverson, W. S., D. M. Waller, and S. L. Solheim. 1988. Forests too deer: Edge effects in northern Wisconsin. *Conserv. Biol.* 2:348–58.

Barrett, G. W. 1968. The effect of an acute insecticide stress on a semi-enclosed grassland ecosystem. *Ecology* 49:1019–35.

Barrett, G. W. 1985. A problem-solving approach to resource management. *BioScience* 35:423–7.

Barrett, G. W. 1987. Applied ecology at Miami University: An integrative approach. *Bull. Ecol. Soc. Am.* 68:154–5.

Barrett, G. W. 1988. Effects of Sevin on small mammal populations in agricultural and old-field ecosystems. *J. Mammal.* 69:731–9.

Barrett, G. W. 1992. Landscape ecology: Designing sustainable agricultural landscapes. *J. Sustainable Agric.* 2:83–103.

Barrett, G. W. 1993. Restoration ecology: Lessons yet to be learned. Pages 113–26 *in* Baldwin, A. D., J. DeLuce, and C. Pletsch (eds.), *Beyond Preservation: Restoring and Inventing Landscapes.* Minneapolis: University of Minnesota Press. 280 pp.

Barrett, G. W., and P. J. Bohlen. 1991. Landscape ecology. Pages 149–61 *in* Hudson, W. E. (ed.), *Landscape Linkages and Biodiversity.* Washington, D.C.: Island Press. 196 pp.

Cole, F. R. 1978. A movement barrier useful in population studies of small mammals. *Am. Midl. Nat.* 100:480–2.

Covey, S. R. 1989. *The Seven Habits of Highly Effective People.* New York: Simon & Schuster. 340 pp.

Crowner, A. W., and G. W. Barrett. 1979. Effects of fire on the small mammal component of an experimental grassland community. *J. Mammal.* 60:803–13.

Diamond, J. 1993. Cougars and corridors. *Nature (London)* 365:16–7.

Fahrig, L., and G. Merriam. 1985. Habitat patch connectivity and population survival. *Ecology* 66:1762–8.

Hall, A. T., P. E. Woods, and G. W. Barrett. 1991. Population dynamics of the meadow vole (*Microtus pennsylvanicus*) in nutrient-enriched old-field communities. *J. Mammal.* 72:332–42.

Hansson, L. 1977. Spatial dynamics of field voles *Microtus agrestis* in heterogeneous landscapes. *Oikos* 29:529–44.

Harper, S. J., E. K. Bollinger, and G. W. Barrett. 1993. Effects of habitat patch shape on population dynamics of meadow voles (*Microtus pennsylvanicus*). *J. Mammal.* 74:1045–55.

Harris, L. D. 1988. Edge effects and conservation of biotic diversity. *Conserv. Biol.* 2:330–2.

Henderson, M. T., G. Merriam, and J. Wegner. 1985. Patchy environments and species survival: Chipmunks in an agricultural mosaic. *Biol. Conserv.* 31:95–105.

Hurlbert, S. H. 1984. Pseudoreplication and the design of ecological field experiments. *Ecol. Monogr.* 54:187–211.

Kemp, J. C., and G. W. Barrett. 1989. Spatial patterning: Impact of uncultivated corridors on arthropod populations within soybean agroecosystems. *Ecology* 70:114–28.

Kozakiewicz, M. 1993. Habitat isolation and ecological barriers: The effect on small mammal populations and communities. *Acta Theriol.* 38:1–30.

La Polla, V. N., and G. W. Barrett. 1993. Effects of corridor width and presence on the population dynamics of the meadow vole (*Microtus pennsylvanicus*). *Landscape Ecol.* 8:25–37.

Laudenslayer, W. F., Jr. 1986. Summary: Predicting effects of habitat patchiness and fragmentation. Pages 331–3 *in* Verner, J., M. L. Morrison, and C. J. Ralph (eds.), *Wildlife 2000: Modeling Habitat Relationships of Terrestrial Vertebrates.* Madison: University of Wisconsin Press. 470 pp.

Lidicker, W. Z., Jr., J. O. Wolff, L. N. Lidicker, and M. H. Smith. 1992. Utilization of a habitat mosaic by cotton rats during a population decline. *Landscape Ecol.* 6:259–68.

Likens, G. E., F. H. Bormann, N. M. Johnson, D. W. Fisher, and R. S. Pierce. 1970. Effects of forest cutting and herbicide treatment on nutrient budgets in the Hubbard Brook watershed ecosystem. *Ecol. Monogr.* 40:23–47.

Likens, G. E., F. H. Bormann, R. S. Pierce, J. S. Eaton, and N. M. Johnson. 1977. *Biogeochemistry of a Forested Ecosystem.* New York: Springer-Verlag. 146 pp.

Lorenz, G. C., and G. W. Barrett. 1990. Influence of simulated landscape corridors on house mouse (*Mus musculus*) dispersal. *Am. Midl. Nat.* 12:348–56.

Lubchenco, J., A. M. Olson, L. B. Brubaker, S. R. Carpenter, M. M. Holland, S. P. Hubbell,

S. A. Levin, J. A. MacMahon, P. A. Matson, J. M. Melillo, H. A. Mooney, C. H. Peterson, H. R. Pulliam, L. A. Real, P. J. Regal, and P. G. Risser. 1991. The Sustainable Biosphere Initiative: An ecological research agenda. *Ecology* 72:371–412.

Maly, M. S., B. A. Knuth, and G. W. Barrett. 1985. Effects of resource partitioning on dispersal behavior of feral house mice. *J. Mammal.* 66:148–53.

Merriam, G., and A. Lanoue. 1990. Corridor use by small mammals: Field measurement for three experimental types of *Peromyscus leucopus. Landscape Ecol.* 4:123–31.

Odum, E. P. 1984. The mesocosm. *BioScience* 34:558–62.

O'Neill, R. V., D. L. DeAngelis, J. B. Wade, and T. F. H. Allen. 1986. *A Hierarchical Concept of Ecosystems.* Princeton, N.J.: Princeton University Press. 253 pp.

Pavuk, D. M., and G. W. Barrett. 1993. Influence of successional and grassy corridors on parasitism of *Plathypena scabra* (F.) (Lepidoptera: Noctuidae) larvae in soybean agroecosystems. *Environ. Entomol.* 22:541–6.

Risser, P. G., J. R. Karr, and R. T. T. Forman. 1984. *Landscape Ecology: Directions and Approaches.* Spec. Publ. 2. Champaign: Illinois Natural History Survey. 18 pp.

Robinson, G. R., R. D. Holt, M. S. Gaines, S. P. Hamburg, M. L. Johnson, H. S. Fitch, and E. A. Martinko. 1992. Diverse and contrasting effects of habitat fragmentation. *Science* 257:524–6.

Stamps, J. A., and M. Buechner. 1985. The territorial defense hypothesis and the ecology of insular vertebrates. *Q. Rev. Biol.* 60:155–81.

Stamps, J. A., M. Buechner, and V. V. Krishnan. 1987a. The effects of edge permeability and habitat geometry on emigration from patches of habitat. *Am. Nat.* 129:533–52.

Stamps, J. A., M. Buechner, and V. V. Krishnan. 1987b. The effects of habitat geometry on territorial defense costs: Intruder pressure in bounded habitats. *Am. Zool.* 27:307–25.

Stueck, K. L., and G. W. Barrett. 1978. Effects of resource partitioning on the population dynamics and energy utilization strategies of feral house mice (*Mus musculus*) populations under experimental field conditions. *Ecology* 59:539–51.

Swihart, R. K., and N. A. Slade. 1984. Road crossing in *Sigmodon hispidus* and *Microtus ochrogaster. J. Mammal.* 65:357–60.

Szacki, J., J. Babińska-Werka, and A. Liro. 1993. The influence of landscape spatial structure on small mammal movements. *Acta Theriol.* 38:113–23.

Temple, S. A., and J. R. Cary. 1988. Modeling dynamics of habitat-interior bird populations in fragmented landscapes. *Conserv. Biol.* 2:340–7.

Turner, M. G. 1989. Landscape ecology: The effect of pattern on process. *Annu. Rev. Ecol. Syst.* 20:171–97.

Urban, D. L., R. V. O'Neill, and H. H. Shugart Jr. 1987. Landscape ecology. *BioScience* 37:119–27.

Wegner, J. F., and G. Merriam. 1979. Movement by birds and small mammals between a wood and adjoining farmland habitats. *J. Appl. Ecol.* 16:349–57.

Wilcove, D. S., C. H. McLellan, and A. P. Dobson. 1985. Habitat fragmentation in the temperate zone. Pages 237–56 *in* M. E. Soulé (ed.), *Conservation Biology: The Science of Scarcity and Diversity.* Sunderland, Mass.: Sinauer Associates. 584 pp.

Williams, C. K., V. A. Witmer, M. Casey, and G. W. Barrett. 1994. Effects of strip-cropping and harvesting on small mammal population dynamics in soybean agroecosystems. *Ohio J. Sci.* 94:94–98.

Yahner, R. H. 1988. Changes in wildlife communities near edges. *Conserv. Biol.* 2:333–9.

9

Population Dynamics of Small Mammals in Fragmented and Continuous Old-Field Habitat

James E. Diffendorfer, Norman A. Slade,
Michael S. Gaines, and Robert D. Holt

Early empirical work pertinent to habitat fragmentation was motivated by island biogeography theory and generally focused on descriptive, community-level patterns of extinctions and species diversity as functions of patch size or isolation (for reviews, see Diamond 1984; Simberloff 1988). Island theory was used to predict the optimal design of reserves; this attempted use of ecological theory has led to continuing debate (Beckon 1993; Gilpin 1988; Quinn and Hastings 1987). However, there is growing recognition that habitat fragmentation acts via ecological processes that affect population dynamics (McCallum 1992; Ostfeld 1992; Pulliam and Danielson 1992) and therefore local abundance (Fahrig and Paloheimo 1988), viability (Fahrig and Merriam 1985; Lande 1987; Roff 1974), and community organization (Holt 1985, 1993). Moreover, ecologists are increasingly becoming aware of how profoundly landscape patterns and mesoscale processes can influence local population dynamics and community structure (Hansson, Chap. 2; Holt 1993; Lidicker, Chap. 1).

As in other areas of ecology, properly designed experiments could be useful in analyzing habitat fragmentation. But in practice such experiments are difficult to design and execute, for several reasons. First, the spatial scale at which habitat fragmentation affects ecological processes is often too large to allow adequate replication or controls in experimental designs (but see Lord and Norton 1990). Second, studies of habitat fragmentation are motivated by conservation issues and so are often done "after the fact." A researcher may not sample a forest before it is cut, but instead must deal with remnants from an already fragmented system. For these reasons, most studies of habitat fragmentation in the past have understandably tended to describe patterns and could only indirectly infer the underlying causal mechanisms. However, ecologists are increasingly using experimental landscapes as model systems for addressing landscape-level questions (Barrett et al., Chap. 8).

Moreover, increasing attention has recently been paid to the demographic mechanisms involved in habitat fragmentation (Barrett et al., Chap. 8; Brown and Ehrlich 1980; Fahrig and Paloheimo 1987; Foster and Gaines 1991; Gaines et al. 1992a,b, 1994; Haila et al. 1993; Harper et al. 1993; Harrison et al. 1988; Henderson et al. 1985; Kareiva 1987, 1990; Laurance, Chap. 3; Lovejoy et al. 1984; Peltonen and Hanski 1991; Soulé et al. 1992; van Apeldoorn et al. 1992; Wegner and Merriam 1979). These studies indicate that habitat fragmentation can have far-reaching and sometimes unexpected effects on populations and communities that are mediated through effects on demographic parameters.

In this chapter, we describe the results of an experimental system in which fragmentation has been imposed on a landscape to investigate the effects of habitat fragmentation on basic population dynamics in old-field habitats in eastern Kansas. Our basic approach is to combine data from two ongoing long-term studies of small-mammal populations in old-field habitats. One study focuses on a continuous area of habitat, the other on a fragmented system (Fig. 9.1). We will compare temporal trends in abundance, survival, sex ratios, and reproduction for three common small-mammal species (*Sigmodon hispidus, Microtus ochrogaster,* and *Peromyscus maniculatus*). Populations in the continuous habitat can be considered "controls," which allows us to make more powerful inferences about the influence of habitat fragmentation on population processes than are permitted by comparisons among patches within the fragmented system (Foster and Gaines 1991; Gaines et al. 1992a,b, 1994). This chapter represents the first attempt to synthesize these long-term data sets to provide a perspective on the overall effects of habitat fragmentation on small-mammal demography.

Earlier analyses from the fragmented site enable us to predict the abundance patterns we may expect when we compare the two sites. The most striking result from the fragmented area study is the clear relationship between body size/block size and density, which has been evident for 7.7 years (Fig. 9.2). Here, "blocks" are defined as spatial units 50 m x 100 m in extent, with clusters of either 15 small patches, six medium patches, or one large patch. The largest-bodied species (*S. hispidus*) has its highest densities on large blocks and is rarely found on blocks of the other sizes. Similarly, the medium-sized species (*M. ochrogaster*) has its highest densities on the blocks with medium-sized patches, and the smallest-bodied species (*P. maniculatus*) has its highest densities on the blocks of small patches. We found

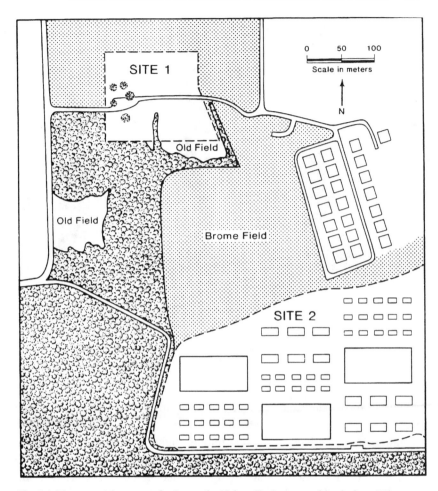

Fig. 9.1. Diagram of the two study sites at the Nelson Environmental Study Area 12 km northeast of Lawrence, Kansas. The continuous area (site 1) is approximately 1.9 ha. The experimentally fragmented system (site 2) has approximately 1.9 ha of successional old-field habitat contained within a total area of 6.9 ha.

this result surprising, because we initially expected all three species to reach their highest densities on the largest blocks. Clearly, the three species have responded to the imposed habitat fragmentation in different ways.

Our working hypothesis for the observed density trends combines consideration of species-specific ecological requirements and interspecific competition (Gaines et al. 1992a,b). We hypothesized that the larger-bodied *S. hispidus* is restricted to the large blocks because it cannot find sufficient

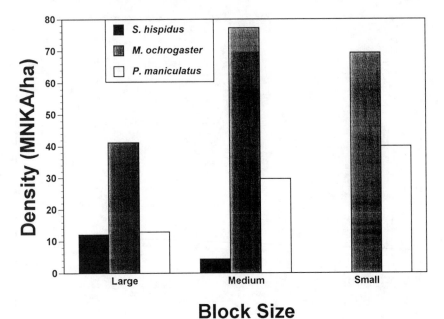

Block Size

Fig. 9.2. Average density (minimum number known alive [MNKA] per hectare) by block size for *Sigmodon hispidus, Microtus ochrogaster,* and *Peromyscus maniculatus.* Data are means of 7.7 years (August 1984 to May 1992) of mark-recapture records on the fragmented study site.

resources to sustain itself on smaller blocks. This habitat restriction in turn provides *M. ochrogaster* with a competitive refuge from *S. hispidus* on the medium and small blocks, thus permitting higher densities there than on large blocks; the likelihood of competition, particularly interference, is suggested from previous studies on the interaction between *S. hispidus* and *M. ochrogaster* (Glass and Slade 1980; Terman 1978).

Competition may explain the patterns observed for *P. maniculatus* as well. *M. ochrogaster* has lower densities on the small blocks than on the medium-sized blocks presumably because, as with *S. hispidus,* the small blocks do not have enough habitat to sustain individuals. *P. maniculatus* may therefore reach its highest densities on small blocks where it can avoid competition with both *S. hispidus* and *M. ochrogaster* (Abramsky et al. 1979; Grant 1971, 1972; Redfield et al. 1977). Furthermore, unlike the other two species, *P. maniculatus* also uses the mowed interstitial areas separating the old-field patches (Foster and Gaines 1991), and thus it may not experience the small amounts of unmowed habitat on the small blocks as isolates.

Given this working hypothesis regarding the underlying processes shaping the densities of the three species on the fragmented area, we made the following species-specific predictions about the abundances on the two study sites being compared: (1) *S. hispidus* should have higher abundances on the continuous site, because it will have more available habitat there than in the fragmented system. (2) *M. ochrogaster* should have higher abundances on the fragmented site, because it provides competitor-free space not available in the continuous area. (3) *P. maniculatus* should have greatest abundances on the fragmented site, because it too may have a competitive refuge there.

Materials and Methods

The study sites are located at the Nelson Environmental Study Area, 12 km northeast of Lawrence, Kansas. The sites at their closest points are no more than 500 m apart and are separated by a brome (*Bromus* sp.) field and low woods (Fig. 9.1). Work on the continuous area was started in 1973 by N. Slade and continues today. This 1.9 ha area is divided by a small dirt road running east-west. The north side of the plot was an abandoned farm field, and the south side was an abandoned pasture and hay field. The southern half of the study site also contains a fencerow with bushes and small trees running north-south.

The fragmentation study began in 1984 when an abandoned agricultural field was disked and allowed to enter secondary succession. Since then, regular mowing between the rectangular habitat patches has created blocks of successional habitat (Fig. 9.1). The habitat patches in the 6.9 ha fragmented site are arranged in three types of blocks. The three large blocks each consist of a single 50 m × 100 m patch. Two medium blocks each include six 12 m × 24 m patches. Of the three small blocks, two consist of 15, and one of 10, 4 m × 8 m patches. The total amount of habitat on the fragmented area is 1.87 ha, nearly identical to the area of old-field habitat on the continuous study site.

Trapping regimes on the two study sites differed. The continuous area contained 98 trap stations, spaced 15 m apart, each with two traps. Traps were checked monthly for three consecutive mornings and the intervening afternoons. The fragmented system contained a total of 287 traps distributed at 267 locations and was trapped twice monthly for two consecutive mornings and the intervening afternoon. Thus, the continuous area had a lower trap density than the fragmented area and was trapped only once a month, compared to twice a month on the fragmented site, although trapping sessions on the continuous area lasted an extra afternoon and morning.

To make comparisons between the two sites, we ignored the difference in trap densities and corrected for the difference in trapping periods. Differences in trap densities could bias our comparisons if traps became saturated when abundances were high; however, even at high numbers, total captures in a trapping session were rarely more than 60% of the possible capture opportunities. Thus, trap densities should have relatively little influence on comparisons between the sites. To make the temporal structure of the two data sets similar, we deleted every other consecutive trapping session from the raw data on the fragmented site and the last afternoon and morning from the raw data on the continuous site. With this trimming, both data sets represent monthly trapping sessions of two mornings and one afternoon.

We have sufficient data on *S. hispidus, M. ochrogaster,* and *P. maniculatus,* the three most abundant species, to make demographic comparisons between the sites. For *S. hispidus* and *M. ochrogaster,* the data begin in the fall of 1984 and continue to the spring of 1991. *P. maniculatus* were not individually marked on the continuous area until the winter of 1989, so we were forced to analyze a smaller data set for this species.

Using Fortran programs developed by C. J. Krebs, we estimated the monthly minimum number known alive (MNKA) and Jolly-Seber survivorship for both entire populations and the resident subsets (residents were defined as individuals captured in at least two monthly trapping periods). Since residents, by definition, survived at least one month, our survival estimates for residents are positively biased but comparable across sites. MNKA estimates could be biased if animals had different probabilities of capture on the two sites. We calculated the probability of capture by dividing the number of animals captured in a trapping session by the MNKA for that session. We also calculated monthly sex ratios and the percentage of animals that were reproductive. Males with descended testes were considered reproductive. Females that were captured with young, were obviously pregnant, or had a clearly open pubic symphysis were considered reproductive. We also classified females as reproductive if they had any two of the following three characteristics: nipples medium or larger, a perforate vagina, or a slightly open pubic symphysis. Transients were counted by subtracting the number of resident animals from the total number of animals present.

All data were analyzed with general linear models in which site, year-season, and sex (depending on the dependent variable) were entered as independent variables (using MINITAB). Treating each three months as a unique block of time (i.e., the year-season variable) focused the analysis on differences between sites while adjusting for seasonal and annual changes in populations. Our procedure was analogous to a paired-comparison *t*-test, but with comparisons being made within each block of time and with sex included as an additional independent variable.

We restricted our analysis to adults when we calculated the percentage of reproductively active individuals, and we considered each sex separately. Adulthood was determined by body mass of at least 60 g for *S. hispidus,* 25 g for *M. ochrogaster,* and 11 g for *P. maniculatus.*

Estimates of the percentages of reproductive adults were weighted by the number of animals handled in a particular trapping period for analysis. Estimates of sex ratios, survivorship, and percentage of transients were weighted by the MNKA estimates for that trapping period. In some cases, data were missing for a particular year-season. If so, we removed that year-season from the analysis and proceeded with a smaller sample size.

Results

Abundances

Abundances of *S. hispidus* for both the total population and the resident subset varied significantly over time (total population, $F = 12.92$; residents, $F = 12.10$; df = 30, 124; $p < 0.001$ in both cases) (Fig. 9.3). Seasonal variation was strong; abundances were highest in the summer and fall and lowest in the winter and spring, and there were some differences among years. Temporal patterns of variation were similar between sites (total population, $r = 0.644$, $n = 31$, $p < 0.001$), but the sites did not maintain the same rank order of abundances through time (significant site by year-season interaction; total population, $F = 2.81$; residents, $F = 3.86$, df = 30, 124; $p < 0.001$ in both cases).

Despite this temporal variation, *S. hispidus* had higher average abundances on the continuous site than on the fragmented site. The means and standard errors are as follows: for the continuous site—total population, 20.27 ± 0.20; residents, 15.06 ± 0.14; for the fragmented site—total population, 14.82 ± 0.20; residents, 9.08 ± 0.14 (total population, $F = 12.10$; residents, $F = 29.9$; df = 1, 124; $p < 0.001$ in both cases). These differences in abundances at the two sites resulted from differences during population peaks (in four years in particular); population sizes were similar on both sites during periods of low numbers (Fig. 9.3a).

Both the total population and the resident subset of *M. ochrogaster* also varied temporally in abundances (total population, $F = 25.09$; residents, $F = 36.56$; df = 30, 124; $p < 0.001$ in both cases) (Fig. 9.4). Unlike *S. hispidus,* however, abundances of *M. ochrogaster* did not show strong annual cycles but did show obvious multiannual cycles; peak abundances were reached on

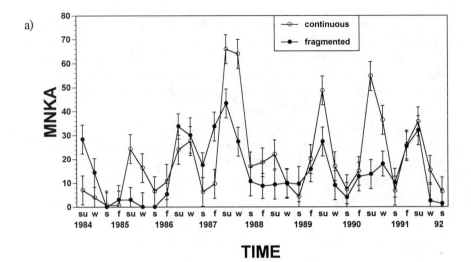

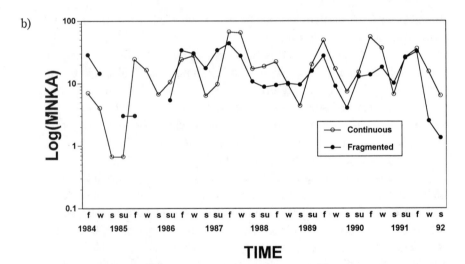

Fig. 9.3. Seasonal average minimum number known alive (MNKA) (a) and log of the seasonal MNKA (b) for *Sigmodon hispidus* on the continuous and fragmented sites from 1984 to 1992. Error bars show the standard errors of the means calculated from the mean squared error from the general linear model.

both sites in the fall of 1987 and again in the winters of 1989 and 1990. These patterns of variation in abundance were similar between sites (total population, $r = 0.74$, $n = 31$, $p < 0.001$), but relative differences in abundances changed through time, leading to a significant site by year-season interaction (total population, $F = 6.66$; residents, $F = 9.57$; df = 30, 124; $p < 0.001$ in both cases). Despite variation in which site had the highest numbers, on average *M. ochrogaster* abundances were twice as high on the fragmented site as on the continuous site. Means and standard errors are as follows: for the continuous site—total population, 40.73 ± 0.39; residents, 32.65 ± 0.29; for the fragmented site—total population, 82.18 ± 0.37; residents, 66.50 ± 0.27 (total population, $F = 192.42$; residents, $F = 228.5$; df = 1, 124; $p < 0.001$ in both cases). In general, *M. ochrogaster* abundances were greater on the fragmented site than on the continuous site but, as with *S. hispidus,* peak abundances differed between sites far more than did low numbers (Fig. 9.4a).

Like the other two species, *P. maniculatus* varied in abundance (both total and resident populations) over time (total population, $F = 14.98$; residents, $F = 24.77$; df = 12, 51; $p < 0.001$ in both cases) (Fig. 9.5). No seasonal trends were evident, although abundances on the fragmented site peaked in 1989. Variation in *P. maniculatus* abundances through time was not correlated between the sites ($r = 0.035$, $p = 0.911$), and there was a significant site by time interaction (total population, $F = 13.47$; residents, $F = 22.28$; df = 12, 51; $p < 0.001$ in both cases). *P. maniculatus* abundances on the fragmented area declined from a peak in 1989; the peak was not observed on the continuous site, which resulted in mean abundances three times higher on the fragmented site than on the continuous site. Means and standard errors are as follows: for the continuous site—total population, 12.15 ± 0.24; residents, 9.14 ± 0.90; for the fragmented site—total population, 36.71 ± 0.24; residents, 27.61 ± 0.90 (total population, $F = 166.64$; residents, $F = 210.15$; df = 1, 51; $p < 0.001$ in both cases). MNKA (y) and number of captures (x) on the continuous site were strongly related ($R^2 = 0.916$; $y = 0.884 + 0.528\ x$; $t = 23.84$; df = 52; $p < 0.001$). We used this regression equation to estimate MNKA from the number of captures prior to 1989, when MNKA estimates were not available for the continuous site. We found that *P. maniculatus* abundances were generally higher on the fragmented site, and not solely as a result of the solitary peak observed on the fragmented site (Fig. 9.5).

Capture Probability

The probability of capture varied over time but not between sites for *S. hispidus* ($F = 4.74$, df = 27, 109, $p < 0.001$) and *M. ochrogaster* ($F = 1.73$,

a)

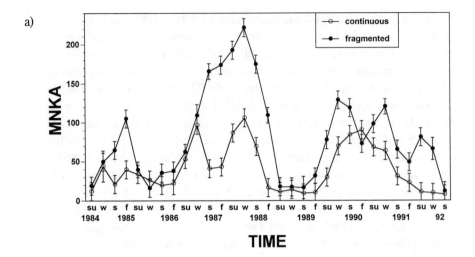

b)

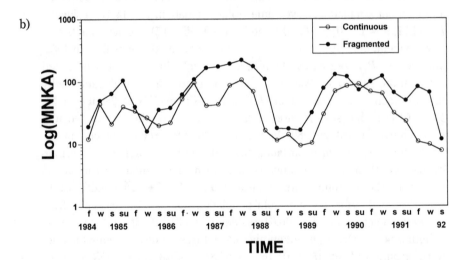

Fig. 9.4. Seasonal average minimum number known alive (MNKA) (a) and log of the seasonal MNKA (b) for *Microtus ochrogaster* on the continuous and fragmented sites from 1984 to 1992. Error bars show the standard errors of the means calculated from the mean squared error from the general linear model.

a)

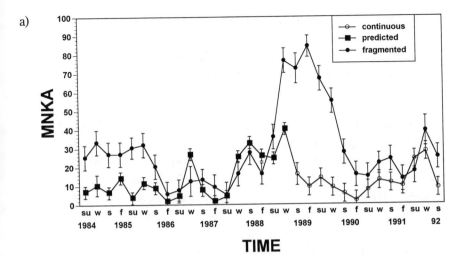

b)

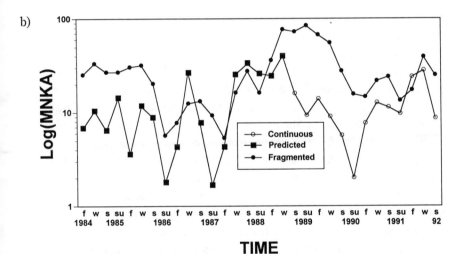

Fig. 9.5. Seasonal average minimum number known alive (MNKA) (a) and log of the seasonal average MNKA (b) for *Peromyscus maniculatus* on the continuous and fragmented sites from 1984 to 1992. Error bars show the standard errors of the means calculated from the mean squared error from the general linear model. The squares on the continuous-site time series are estimates from a regression of MNKA on the number of captures.

df = 30, 124, p = 0.019). For *P. maniculatus,* the average probability of capture was higher on the continuous site (0.93 ± 0.034) than on the fragmented site (0.75 ± 0.019) (F = 20.33, df = 1, 50, p < 0.001), a trend opposite to that expected if capture probability affects abundance.

Survivorship

We estimated survivorship for both the entire population and the resident population. *S. hispidus* survivorship for the total population did not differ by site or sex but changed through time, i.e., year-season (F = 1.48, df = 26, 200, p = 0.071). Because year-season was the only significant independent variable in the model, we tested whether season alone could account for the temporal variation in survivorship by comparing residual sums of squares from a simple model with only season as the independent variable with those from a more complex model with year-season as the independent variable (Draper and Smith 1981). The complex model produced a significant increase in the model's fit (F = 2.08, df = 23, 281, p = 0.003). Hence, both year and season were statistically important in explaining the variation in survivorship.

Survivorship of the total population was highest in spring and lowest in summer (Table 9.1). Resident *S. hispidus* survivorship varied only with year-season (F = 2.11, df = 24, 261, p = 0.002). Unlike total *S. hispidus* survivorship, there was no difference in the amount of variation in resident survivorship explained by the model with year-season as the independent variable compared to the model with the season variable alone, indicating that season explained most of the variation in resident survivorship. Survivorship of residents varied seasonally (F = 5.58, df = 3, 282, p < 0.001) but in different ways than for the total population (Table 9.1). Unlike the total population, resident survivorship was highest in the fall and lowest in the winter.

As with *S. hispidus,* survivorship for the total *M. ochrogaster* population varied only over time (F = 3.38, df = 30, 238, p < 0.001). Furthermore, a model with year-season as the independent variable fit the data significantly better than a model with just season (F = 3.640, df = 27, 331, p < 0.001). Thus, *M. ochrogaster* survivorship varied with both year and season. On average, survivorship was highest in the fall and winter and lowest in the spring and summer (Table 9.1). Resident *M. ochrogaster* survivorship also varied with time (F = 3.66, df = 30, 235, p < 0.001) and, as with the total population, a model with year-season fit the data better than a model with season alone (F = 3.33, df = 27, 328, p < 0.001). Resident survivorship was

Table 9.1. 28-day Jolly-Seber survivorships (mean ± one standard error) by season for total and resident populations of *Sigmodon hispidus, Microtus ochrogaster,* and *Peromyscus maniculatus*

Species Season	Total population	N	Resident population	N
S. hispidus				
Fall	.582 ± .022	98	.805 ± .021	84
Winter	.582 ± .029	78	.685 ± .023	78
Spring	.696 ± .029	60	.714 ± .038	60
Summer	.472 ± .049	86	.786 ± .035	74
M. ochrogaster				
Fall	.744 ± .020	98	.856 ± .017	98
Winter	.767 ± .018	90	.861 ± .015	90
Spring	.676 ± .020	88	.759 ± .017	88
Summer	.641 ± .021	86	.787 ± .019	86
P. maniculatus				
Fall	.639 ± .036	36	.798 ± .032	36
Winter	.723 ± .057	36	.824 ± .030	36
Spring	.454 ± .041	36	.604 ± .033	36
Summer	.663 ± .047	38	.856 ± .030	26

also highest in the fall and winter and lowest in the spring and summer (Table 9.1).

P. maniculatus survivorship also varied over time ($F = 3.37$, df = 12, 116, $p < 0.001$). However, total *P. maniculatus* populations had higher average survivorship on the fragmented site (0.62 ± 0.022) than on the continuous site (0.51 ± 0.041) ($F = 5.09$, df = 1, 116, $p = 0.026$). To determine whether year-season was a better predictor of survivorship than season alone, we ran a model including site and either season or year-season. The overall fit to the data did not differ between the two models, indicating that season explained the variation in survivorship as well as did year-season. Average survivorship for the total population varied in similar patterns on both grids, with highest survivorship in the winter and lowest in the spring. Spring survivorship on the continuous site was 1.5 times lower than on the fragmented site, which led to a significant difference in survivorship between the grids (Table 9.1). Survivorship of residents did not differ between sites. As with the other two species, resident survivorship varied with time ($F = 3.07$, df = 11, 84, $p = 0.002$). Resident survivorship changed with season, and year-season was a better predictor of survivorship than season ($F = 2.786$, df = 8, 120, $p = 0.007$). Residents had highest average survivorship in the summer and lowest in the spring (Table 9.1).

Table 9.2. Percentage of male *Sigmodon hispidus, Microtus ochrogaster,* and *Peromyscus paniculatus* in the continuous and fragmented study sites

	Study Site			
Species	Continuous	N^a	Fragmented	N^a
S. hispidus	.573 ± .017	1,867	.450 ± .016	1,433
M. ochrogaster	.500 ± .009	3,678	.443 ± .006	7,979
P. maniculatus	.473 ± .021	458	.519 ± .012	1,426

[a]Since the general linear models were weighted by the minimum number known alive (MNKA), N represents the total MNKA summed over all trapping sessions (mean ± one standard error).

Sex Ratios

S. hispidus sex ratio varied over time ($F = 1.60$, df = 27, 109, $p = 0.047$) (Fig. 9.6a) but in different patterns at each site, leading to a significant site by year-season interaction ($F = 3.04$, df = 27, 109, $p < 0.001$). On average, *S. hispidus* populations in continuous habitat were male-biased compared to those in the fragmented habitat ($F = 27.99$, df = 1, 109, $p < 0.001$) (Table 9.2). In *M. ochrogaster* populations, sex ratio varied with year-season ($F = 2.62$, df = 30, 124, $p < 0.001$) (Fig. 9.6b) and in different ways on each site ($F = 2.49$, df = 30, 124, $p < 0.001$). Populations in the fragmented habitat were on average female-biased ($F = 27.92$, df = 1, 124, $p < 0.001$) (Table 9.2). In *P. maniculatus* populations, sex ratio varied in different patterns on each site over time, resulting in a significant site by year-season interaction ($F = 2.37$, df = 12, 50, $p = 0.017$).

Reproduction

The proportions of adult female *S. hispidus* and *M. ochrogaster* that were judged to be reproductive changed over time (*S. hispidus*, $F = 3.14$, df = 24, 87, $p < 0.001$; *M. ochrogaster*, $F = 4.37$, df = 30, 121, $p < 0.001$). Higher proportions of female *S. hispidus* and *M. ochrogaster* were reproductive on the continuous habitat than on the fragmented habitat (*S. hispidus*, $F = 3.69$, df = 1, 87, $p = 0.058$; *M. ochrogaster*, $F = 9.60$, df = 1, 121, $p < 0.001$) (Table 9.3). The proportion of reproductive *S. hispidus* males varied with time ($F = 4.23$, df = 25, 92, $p < 0.001$). For male *M. ochrogaster,* the proportion varied over time and in different patterns on each site ($F = 2.39$, df = 30, 119, $p < 0.001$). The average proportions of reproductive male *S. hispidus* and *M. ochrogaster* were similar between sites. The proportion of reproductive male *P. maniculatus* changed with time ($F = 4.37$, df = 12, 49, $p < 0.001$), but on average, greater

a)

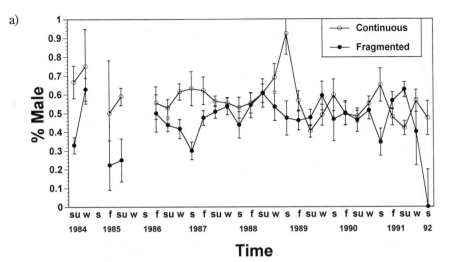

b)

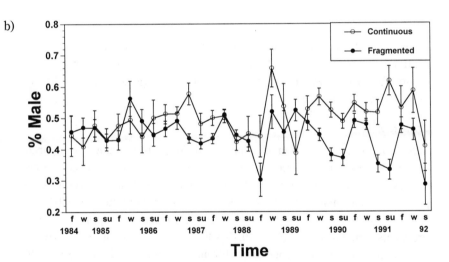

Fig. 9.6. Seasonal sex ratio estimates for *Sigmodon hispidus* (a) and *Microtus ochrogaster* (b) on the continuous and fragmented sites from 1984 to 1992. Error bars show the standard errors of the means calculated from the mean squared error from the general linear model.

Table 9.3. Percentage of reproductively active adult male and female *Sigmodon hispidus,*
Microtus ochrogaster, and *Peromyscus maniculatus* trapped in continuous and fragmented
sites

Species Site	Male	N^a	Female	N^a
S. hispidus				
Continuous	.425 ± .048	565	.399 ± .055	478
Fragmented	.431 ± .059	316	.216 ± .046	366
M. ochrogaster				
Continuous	.741 ± .030	1,224	.637 ± .035	1,191
Fragmented	.787 ± .019	2,250	.510 ± .022	2,820
P. maniculatus				
Continuous	.486 ± .058	163	.442 ± .058	159
Fragmented	.638 ± .037	437	.336 ± .029	424

[a]Since general linear models were weighted by the number of adult animals captured in a
trapping session, N represents the total number of adults captured (mean ± one standard
error).

proportions of males were reproductive in the fragmented habitat than in the
continuous habitat ($F = 4.83$, df = 1, 49, $p = 0.003$) (Table 9.3). The proportion
of reproductive *P. maniculatus* females varied with time ($F = 6.16$, df = 12, 45,
$p < 0.001$) but not by site.

Transients

In *S. hispidus,* the proportion of transient individuals (those animals captured
only in a single month) varied over time ($F = 6.69$, df = 26, 108, $p < 0.001$). However, populations in the fragmented area tended to have a higher average proportion of transients ($0.346 ± 0.026$) than those on the continuous site ($0.287 ±$
0.023) ($F = 2.98$, df = 1, 108, $p = 0.087$). In *M. ochrogaster,* the proportion of
transients varied with time ($F = 4.16$, df = 30, 108, $p < 0.001$) but not by site.

In *P. maniculatus,* the proportion of transients varied with time ($F = 2.60$,
df = 12, 50, $p = 0.009$), and the pattern of variation differed among the sites
($F = 2.27$, df = 50, 75, $p = 0.021$). The proportion of transients remained fairly
constant on the fragmented site, fluctuating from 0.18 to 0.42. On the continuous site, the proportion transient showed seasonal trends, decreasing every
winter and then rising to high levels in spring or fall. Transience on the continuous site ranged from 0.07 to 0.67. Although the average proportion of
transients was the same for the two sites (0.29), *P. maniculatus* populations
on the continuous area had more than three times the temporal variance in
the proportion of transients as populations on the fragmented site (continu-

ous, s^2 = 0.00027; fragmented, s^2 = 0.00085; one-tailed F = 3.148, df = 12, 12, p = 0.028).

Discussion

Fragmentation influenced species abundances in the manner we predicted when we projected from the spatial patterns in abundance within the fragmented area alone. *S. hispidus,* the largest-bodied species, had higher abundances in the continuous area, whereas the two smaller-bodied species had higher abundances in the fragmented system.

The critical question is, by what demographic processes has habitat fragmentation influenced these differences in abundance? In *S. hispidus,* abundance differences are attributable to the large differences in peak abundances between sites. *M. ochrogaster* had consistently higher abundances, particularly peak abundances, on the fragmented site. *P. maniculatus* had consistently higher abundances on the fragmented site, and densities were less variable than those for the other species. These findings are intriguing. We can use the other demographic data presented to distinguish among plausible processes by which habitat fragmentation influences population dynamics.

If the observed abundance differences were primarily due to differences in the amount of available habitat, disentangled from other ecological processes, then the demographic variables we measured should have been equal between sites. However, our data revealed species-specific differences between the sites in demographic variables, despite similar total amounts of habitat. Apparently, habitat fragmentation leads to species-specific demographic responses, independent of simple area effects.

Sigmodon hispidus

Abundance differences between the sites for *S. hispidus* resulted primarily from higher peaks on the continuous site. These higher peaks indicate times of higher per capita growth rates on the continuous site. We infer higher growth rates from a plot of log MNKA versus time, which shows a consistently steeper slope on the continuous site as populations climb from seasonal lows (Fig. 9.3b).

A number of distinct demographic processes could cause the observed site differences in per capita growth rates for *S. hispidus.* First, higher birth rates on the continuous grid could result in higher growth rates and thus abundances. A higher proportion of females were reproductive on the continuous site. The restriction of activity to the three large blocks in the fragmented site may prevent access to the level of resources needed to sustain high reproductive rates. Furthermore, medium and small patches may be too small to support reproduction. However, abundances of females on smaller patches are too low to definitively test this hypothesis.

Second, movements and social interactions could influence abundances. The continuous site could experience more immigration or less emigration than the fragmented site. The higher percentage of transients on the fragmented site could result from differences either in survival or in dispersal. Since survival rates of residents were similar across the sites, we assume that the differences in transience indicate that fragmentation likely influences movement behavior. Patch geometry could also influence the consequences of intraspecific interference: the greater edge-to-interior ratios on the fragmented site might intensify territorial behavior, because edge provides territorial boundaries that are defensible at low costs, permitting more effective defense of nonedge borders. This could reduce the number of territories per unit area. If so, then subordinate animals would be more prone to leave the fragmented site as conditions became crowded, and potential immigrants would have a more difficult time establishing in patches. The trend toward higher percentages of transients in the fragmented area supports this scenario.

Third, abundances on the fragmented site may not reach the high values seen on the continuous site simply because there is less available habitat overall. Presumably because of its large body size, *S. hispidus* cannot use medium or small habitat blocks for long-term survival and reproduction (see Gaines et al. 1992a,b, 1994). Excluding the medium and small blocks, the fragmented site has 22.7% less habitat available than the continuous area. Average total abundances were 26.9% less on the fragmented site than on the continuous area, which suggests that the decrease in usable habitat may account for much of the difference between sites. However, average resident *S. hispidus* abundances were nearly 40% lower on the fragmented area. The difference between total and resident populations reflects the greater transience observed on the fragmented site and further indicates that habitat fragmentation reduces numbers more than can be accounted for merely by the reduction in suitable area.

In summary, habitat fragmentation reduces *S. hispidus* populations in three distinct demographic ways. First, it lowers the proportion of reproductively active females, resulting in higher per capita growth rates and thus higher maximal abundances in favorable years on the continuous site. Second, a greater proportion of animals are transient on the fragmented system, indicating that animals leave faster or stay shorter amounts of time there.

Third, because few individuals can occupy the smaller blocks of habitat, there is less total habitat available on the fragmented site.

Microtus ochrogaster

We hypothesized that competitor-free space on the fragmented site should lead to higher abundances of *M. ochrogaster* there than on the continuous site. The average abundance of *S. hispidus* on medium blocks is one (1.29 ± 0.25) per block, and this species has been captured on the small patches only 33 times in 7.7 years. Medium and small blocks represent 22.7% of the available habitat on the fragmented site, yet they account for 42.6% of the average *M. ochrogaster* abundance. Thus, in the absence of *S. hispidus, M. ochrogaster* seems capable of achieving higher densities on the smaller blocks (Gaines et al. 1994). This block-specific effect makes a significant contribution to the overall population size of this species on the fragmented site.

As with *S. hispidus,* some of the differences between sites for *M. ochrogaster* resulted from high peak abundances on the fragmented site that did not occur on the continuous site. However, unlike with *S. hispidus,* plots of log MNKA versus time indicate that per capita growth rates during periods when *M. ochrogaster* was rebounding from population lows were similar between the grids and hence do not explain the differences in peak abundances we observed between the sites (Fig. 9.4b). Thus, we must find other explanations for the peak abundance differences.

If for a given series of trapping periods, abundances are, on average, higher on the fragmented site, and if per capita growth rates are similar in both sites, then the population on the fragmented site will grow to higher levels. This effect may have occurred in our system, but in many cases the low pregrowth abundances were roughly equal in both sites. It seems more likely that the differences in abundance reflect differences in "carrying capacity" rather than in maximal growth rates.

Our findings that adult sex ratios are female-biased on the fragmented site and that the proportion of reproductive females is higher on the continuous site merit further discussion. Female-biased sex ratios may reduce the proportion of reproductive females if *M. ochrogaster* is monogamous (Carter and Getz 1993). Movement between blocks is male-biased (J. E. Diffendorfer, unpub. data). If dispersal is correlated with these movements, then fragmentation may increase male dispersal and thus skew the sex ratio toward females. It seems doubtful that this effect could fully explain the differences in mean abundance, because the percentage of transients does not differ between the grids.

Per capita female reproduction may be greater on the continuous site for reasons similar to those suggested above for *S. hispidus.* Small and medium-sized patches may not be ideal breeding habitat for *M. ochrogaster.* Further-more, the higher edge-to-interior ratios on the fragmented site may indirectly reduce the number of female territories by sharpening the effectiveness of interference and thus reduce reproductive activity, particularly for subordi-nate individuals.

In summary, the per capita population growth rates of *M. ochrogaster* when rebounding from low numbers are similar on fragmented and continuous sites. Yet these voles reach higher abundances on the fragmented site be-cause they ultimately achieve higher peak densities there, we suspect mainly because 23% of the fragmented site (i.e., the medium and small patches) is competitor-free. Testing this hypothesis will require further work.

Peromyscus maniculatus

P. maniculatus showed the greatest positive response to fragmentation, with average abundances nearly three times higher on the fragmented site. How-ever, this may be misleading, because more than 50% of the average MNKA can come from the interstitial or matrix area (Foster and Gaines 1991). Including the interstitial area, the fragmented site contains approximately 6.9 ha of mowed and successional vegetation. Extrapolating total abundance in the continuous habitat (1.9 ha) to an area of 6.9 ha results in an abundance of 44, slightly higher than observed on the fragmented site. However, we only set traps in 1.87 ha of the 6.9 ha on the fragmented site. Thus we may have underestimated density for the fragmented site because we had no traps in nearly three-quarters of the habitat available. If we use the finding of Foster and Gaines (1991) that about 50% of the *P. maniculatus* population was found in the interstitial matrix areas, then the overall abundances of *P. manic-ulatus* on the entire fragmented site (36.6 x 2 = 73.2) would be higher than the abundance on a continuous site of equal size (44.1). Thus, *P. maniculatus* abundances on the fragmented area appear to be higher than the abun-dances on a continuous site regardless of how we interpret the data.

The higher abundances of *P. maniculatus* on the fragmented site may re-sult from the abundant food supply in the interstitial area. Furthermore, *P. maniculatus* faces no competitors in the interstitial area and hence, like *M. ochrogaster,* has a huge area of competitor-free space not found on the continuous site.

The increased habitat area and reduced competition are reflected, we sug-gest, in higher average survivorship and higher percentage of reproductive males for the total *P. maniculatus* population on the fragmented site compared to the continuous site. However, despite the higher survivorship and male reproductive activity, a plot of log MNKA versus time gave no evidence of higher per capita growth rates on the fragmented site. The higher percentage of reproductive males on the fragmented site may result from the general increase in resources available in the interstitial areas.

P. maniculatus was the only species for which the probability of capture differed between sites. The lower probability of capture on the fragmented grid fits our scenario of many individuals spread out over the entire 6.9 ha fragmented site, with traps covering only 1.87 ha. Our results agree with those of Sietman et al. (1994), who found that *P. maniculatus* was common in old-field habitats, native tallgrass prairie, and mowed hay fields. This species probably views our system as a complex of two major habitat types, namely, old-field patches interspersed in a "sea" of lower-quality "turf" habitat.

The percentage of transients fell every winter and rose in the spring and summer on the continuous site but not on the fragmented site. This suggests that larger, continuous areas may be better overwintering habitat for *P. maniculatus.* If so, then the percentage of transients should be lower on large blocks than on smaller blocks in the fragmented site. Although we have not analyzed percentage of transience by block size, persistence rates appear to be higher on large blocks than on smaller blocks (Gaines et al. 1994), indicating that animals stay longer on large blocks. Thus, the smaller patches of habitat in the fragmented area may have higher transience than large blocks.

An important direction for future work in the fragmented system is analysis of how the demographic effects documented here and elsewhere (Foster and Gaines 1991; Gaines et al. 1992a,b, 1994) reflect underlying processes such as resource availability, social interactions (R. Pierotti and P. Wilson, pers. comm.), individual movement patterns (J. E. Diffendorfer, unpub. data), spacing behavior (Kozakiewicz and Szacki, Chap. 5), and risk of predation (Oksanen and Schneider, Chap. 7)—all of which are likely to vary as a function of the degree of habitat fragmentation (Hansson, Chap. 2).

Conclusions

Our data show that species respond differently to fragmentation, not just in overall abundance, but also in detailed demographic parameters. For *S. hispidus,* fragmentation reduced habitat availability and was associated with fewer reproductively active females and an increase in transience. These factors collectively influence population growth rates and ultimately the magnitude of peaks in abundances. For *M. ochrogaster,* fragmentation seems to lower female reproductive rates, but this is likely offset by an easing of competitive pressure from *S. hispidus;* the net result is that *M. ochrogaster* abundances are higher on the fragmented site. For *P. maniculatus,* habitat fragmentation actually creates suitable competitor-free habitat, which increases survivorship and reproductive activity, resulting in a higher *P. maniculatus* population in the fragmented area.

Our comparison of continuous and fragmented habitats raises a number of interesting unanswered questions. For example, the differences between the sites in sex ratios of *S. hispidus* and *M. ochrogaster* and in the percentage of reproductive adults (for at least one sex) in all three species indicate underlying variation between the sites that we cannot yet explain. Designing experiments and field protocols to interpret these patterns will be difficult. However, understanding individual demographic responses is essential to deepening our understanding of habitat fragmentation. Moreover, our analyses have focused on only a single, rather coarse landscape attribute—patch size. Other attributes, such as patch configuration and landscape arrangement, are likely to have significant effects on total population responses, reflecting spatial influences on individual demographic parameters (Barrett et al., Chap. 8).

Doak et al. (1992, 332) observed that "few theoretical or empirical studies treat fragmentation in a way that would satisfy the concerns most conservation biologists have about habitat fragmentation." We feel more can be gleaned from experimental model approaches in studies of habitat fragmentation. The contributions by Barrett et al. (Chap. 8) and Laurance (Chap. 3) reinforce this conclusion, albeit from different perspectives. Experimental approaches are necessarily conducted at modest spatial scales, relative to those of concern for conservation biology and landscape ecology (Lidicker, Chap. 1). Nonetheless, experimental systems can sharpen the descriptive, pattern-oriented, and modeling approaches usually required for broad, landscape-level analyses.

Summary

This chapter describes how habitat fragmentation influences temporal trends in abundance and an array of demographic parameters for three small-mammal species, *S. hispidus, M. ochrogaster,* and *P. maniculatus.* From August 1984 to May 1992, we collected capture-recapture data on a single continuous area spanning 1.9 ha of successional old-field habitat and on patches of successional old-field habitat of varying size but totaling 1.87 ha. Numbers of each species varied over time and between sites. MNKAs in the continuous and fragmented areas averaged 20.3 and 14.8, respectively, for *S. hispidus,* 40.7 and 82.2 for *M. ochrogaster,* and 12.2 and 36.7 for *P. maniculatus;* the differences between the two sites in mean abundances were statistically significant ($p < 0.05$) for all three species. Survival rates varied with season for all species but were different between areas ($p < 0.10$) only for *P. maniculatus,* which showed higher survival on the fragmented area. Survival among resident individuals was less variable; it was not significantly different in the two areas ($p > 0.30$); and it was about 15% greater than survival among the total populations (including transients). Transients made up 32% of *S. hispidus,* 23% of *M. ochrogaster,* and 29% of *P. maniculatus* populations regardless of habitat; thus, all abundance differences were caused by differences in abundances of resident animals. *Sigmodon* and *Microtus* males were proportionally more common on the continuous site. Greater proportions of *S. hispidus* and *M. ochrogaster* females were reproductive in continuous habitat, whereas male *Peromyscus* were more likely to be scrotal in the fragmented area.

Our results are consistent with the idea that fragmentation has distinct effects on each species. Fragmentation reduced the amount of habitat available to *S. hispidus* and thereby may offer *Microtus* a competitive refuge from *Sigmodon. P. maniculatus* may be attracted to the closely mowed and competitor-free interstices between habitat patches and hence in effect inhabit an area much larger than the sum of our old-field patches. Our results underscore the value of experimental field studies in analyzing the processes influenced by habitat fragmentation.

Acknowledgments

We thank Bill Lidicker for helpful comments on the manuscript and Charles Krebs for the use of his Fortran programs. We are indebted to the numerous undergraduate and graduate students who assisted with the trapping and data entry. We acknowledge the financial support of the University of Kansas (General Research Fund and Experimental and Applied Ecology Program) and the National Science Foundation (BSR-8718089 and DEB-9308065).

Literature Cited

Abramsky, Z., M. I. Dyer, and D. Harrison. 1979. Competition among small mammals in experimentally perturbed areas of shortgrass prairie. *Ecology* 60:530–6.

Beckon, W. N. 1993. The effect of insularity on the diversity of land birds in the Fiji Islands: Implications for refuge design. *Oecologia* 94:318–29.

Brown, I. L., and P. R. Ehrlich. 1980. Population biology of the checkerspot butterfly, *Euphydryas chalcedona,* structure of the Jasper Ridge colony. *Oecologia* 47:239–51.

Carter, S. C., and L. L. Getz. 1993. Monogamy and the prairie vole. *Sci. Am.* 268:100–7.

Diamond, J. M. 1984. "Normal" extinctions of isolated populations. Pages 191–246 *in* Nitecki, M. H. (ed.), *Extinctions.* Chicago: University of Chicago Press. 354 pp.

Doak, E. D., P. C. Marino, and P. M. Kareiva. 1992. Spatial scale mediates the influence of habitat fragmentation on dispersal success: Implications for conservation. *Theor. Popul. Biol.* 41:315–36.

Draper, N. M., and H. Smith. 1981. *Applied Regression Analysis.* New York: John Wiley & Sons. 709 pp.

Fahrig, L., and G. Merriam. 1985. Habitat patch connectivity and population survival. *Ecology* 66:1762–8.

Fahrig, L., and J. E. Paloheimo. 1987. Interpatch dispersal of the cabbage butterfly. *Can. J. Zool.* 65:616–22.

Fahrig, L., and J. E. Paloheimo. 1988. Determinants of local population size in patchy habitats. *Theor. Popul. Biol.* 34:194–213.

Foster, J., and M. S. Gaines. 1991. The effects of a successional habitat mosaic on a small mammal community. *Ecology* 72:1358–73.

Gaines, M. S., J. E. Diffendorfer, J. Foster, and F. P. Wray. 1994. The effects of habitat fragmentation on three species of small mammals in eastern Kansas. *Pol. Ecol. Stud.* 20:159–71.

Gaines, M. S., J. Foster, J. E. Diffendorfer, W. E. Sera, R. D. Holt, and G. R. Robinson. 1992a. Population process and biological diversity. *Trans. North Am. Wildl. Nat. Resour. Conf.* 57:252–62.

Gaines, M. S., G. R. Robinson, J. E. Diffendorfer, R. D. Holt, and M. L. Johnson. 1992b. The effects of habitat fragmentation on small mammal populations. Pages 875–85 *in* McCullough, R. D., and R. H. Barrett (eds.), *Wildlife 2001: Populations.* London: Elsevier Applied Science. 1,163 pp.

Gilpin, M. E. 1988. A comment on Quinn and Hastings: Extinction in subdivided habitats. *Conserv. Biol.* 2:290–2.

Glass, G. E., and N. A. Slade. 1980. The effect of *Sigmodon hispidus* on spatial and temporal activity of *Microtus ochrogaster:* Evidence for competition. *Ecology* 6:358–70.

Grant, P. R. 1971. Experimental studies of competitive interaction in a two-species system. III. *Microtus* and *Peromyscus* species in enclosures. *J. Anim. Ecol.* 40:323–50.

Grant, P. R. 1972. Interspecific competition among rodents. *Annu. Rev. Ecol. Syst.* 3:79–106.

Haila, Y., I. K. Hanski, and S. Raivio. 1993. Turnover of breeding birds in small forest fragments: The "sampling" colonization hypothesis corroborated. *Ecology* 74:714–25.

Harper, S. J., E. K. Bollinger, and G. W. Barrett. 1993. The effects of habitat patch shape on population dynamics of meadow voles (*Microtus pennsylvanicus*). *J. Mammal.* 74:1045–55.

Harrison, S., D. D. Murphy, and P. R. Ehrlich. 1988. Distribution of the bay checkerspot butterfly, *Euphydryas editha bayensis:* Evidence for a metapopulation model. *Am. Nat.* 132:360–82.

Henderson, M. T., G. Merriam, and J. Wegner. 1985. Patchy environments and species survival: Chipmunks in an agricultural mosaic. *Biol. Conserv.* 31:95–105.

Holt, R. D. 1985. Population dynamics in two-patch environments: Some anomalous consequences of an optimal habitat distribution. *Theor. Popul. Biol.* 28:181–208.

Holt, R. D. 1993. Ecology at the mesoscale: The influence of regional processes on local communities. Pages 77–88 *in* Ricklefs, R., and D. Schluter (eds.), *Species Diversity in Ecological Communities: Historical and Geographical Perspectives.* Chicago: University of Chicago Press. 414 pp.

SMALL-MAMMAL POPULATION DYNAMICS IN OLD FIELDS   199

Kareiva, P. M. 1987. Habitat fragmentation and the stability of predator-prey interactions. *Nature (London)* 326:388–90.

Kareiva, P. M. 1990. Population dynamics in spatially complex environments: Theory and data. *Philos. Trans. R. Soc. London B* 330:175–90.

Lande, R. 1987. Extinction thresholds in demographic models of territorial populations. *Am. Nat.* 130:624–35.

Lord, J. M., and D. A. Norton. 1990. Scale and the spatial context of fragmentation. *Conserv. Biol.* 4:197–202.

Lovejoy, T. E., J. M. Rankin, R. O. Bierregaard Jr., K. S. Brown, L. H. Emmons, and M. E. Van der Voort. 1984. Ecosystem decay of Amazon forest remnants. Pages 295–325 *in* Nitecki, M. H. (ed.), *Extinctions*. Chicago: University of Chicago Press. 354 pp.

McCallum, H. I. 1992. Effects of immigration on chaotic population dynamics. *J. Theor. Biol.* 154:277–84.

Ostfeld, R. S. 1992. Effects of habitat patchiness on population dynamics: A modelling approach. Pages 851–63 *in* McCullough, D. R., and R. H. Barrett (eds.), *Wildlife 2001: Populations*. London: Elsevier Applied Science. 1,163 pp.

Peltonen, A., and I. Hanski. 1991. Patterns of island occupancy explained by colonization and extinction rates in shrews. *Ecology* 72:1698–708.

Pulliam, H. R., and B. J. Danielson. 1992. Sources, sinks, and habitat selection: A landscape perspective on population dynamics. *Am. Nat.* (*Suppl.*) 137:S50–66.

Quinn, J. F., and A. Hastings. 1987. Extinction in subdivided habitats. *Conserv. Biol.* 1:198–208.

Redfield, J. A., C. J. Krebs, and M. J. Taitt. 1977. Competition between *Peromyscus maniculatus* and *Microtus townsendii* in grasslands of coastal British Columbia. *J. Anim. Ecol.* 46:607–16.

Roff, D. A. 1974. Spatial heterogeneity and the persistence of populations. *Oecologia* 15:245–58.

Sietman, B. E., W. B. Fothergill, and E. J. Finck. 1994. Effects of haying and old-field succession on small mammals in tallgrass prairie. *Am. Midl. Nat.* 131:1–18.

Simberloff, D. 1988. The contribution of population and community biology to conservation science. *Annu. Rev. Ecol. Syst.* 19:473–511.

Soulé, M. E., A. C. Alberts, and D. T. Bolger. 1992. The effects of habitat fragmentation on chaparral plants and vertebrates. *Oikos* 63:39–47.

Terman, M. R. 1978. Population dynamics of *Microtus ochrogaster* and *Sigmodon hispidus* in central Kansas. *Trans. Kans. Acad. Sci.* 77:337–51.

van Apeldoorn, R. C., W. T. Oostenbrink, A. van Winden, and F. F. van der Zee. 1992. Effects of habitat fragmentation on the bank vole, *Clethrionomys glareolus,* in an agricultural landscape. *Oikos* 65:265–74.

Wegner, J. F., and G. Merriam. 1979. Movements of birds and small mammals between a wood and adjoining farmland habitats. *J. Appl. Ecol.* 16:349–57.

Epilogue

Have we learned anything important, have we enjoyed our task, and have we provided anything significant for the future? Those seem to be the obvious questions that confront us as we complete our work on this project.

How much we have gained intellectually will probably be in direct proportion to the number of these nine chapters we have read, absorbed, and tried to make a part of ourselves. I suspect that all of us have gained from our participation and from our collaboration. For most of us, this project began in Sydney and has continued through the exchange of materials and ideas during the writing process. Publication should further strengthen these intellectual connections and synergisms. I hope our readers will also find what we have to say intellectually profitable. Certainly, we have tried to be interesting, to bring out differences of opinion, and to push ourselves to new levels of understanding.

Have we experienced our efforts as satisfying and enjoyable? No doubt for all of us, this project has stressed to the utmost our already full agendas. Still, it is satisfying to achieve collaboration across three continents, and with so many colleagues representing a great diversity of backgrounds, approaches, and experiences. Personally, I find it rewarding to assist colleagues for whom English is not their native language to achieve a better level of communication. Expressing complicated ideas is difficult enough in one's native tongue. If science is to be truly world-encompassing, all scientists need to be willing to put some effort into helping each other. Friendships are rewarding, especially when they can transcend disagreements; synergism is a delight.

Finally, and most important, what have we achieved that may be significant for the future? At least we have improved our own intellects, and we hope those of our readers. We have assembled and synthesized a large amount of literature and personal experiences bearing on landscape aspects of mammalian ecology and conservation. We have exposed some important controversies and pointed out some promising directions for future investigation. But, can we claim more? I think we can. A landscape perspective, supported by whatever conceptual nuances we prefer, is certainly a wave of the future, for theoretical ecology as well as for conservation biology. In fact, the authors of the chapters in this book are unanimous in their belief in the critical importance of landscape ecology to conservation. And, as is becoming increasingly apparent, our successes or failures in conserving this planet's biodiversity will have a strong impact on our own species' future.

We hope this book will be a useful foundation for the explosive growth in landscape ecology that has already begun and that it will be a valuable resource even for those not oriented toward mammals. As we have pointed out, mammals are often keystone and/or indicator species in their communities, and hence our efforts to conserve them will have beneficial corollary effects on their entire communities.

It is in the conceptual arena, however, that we most aspire to provide important grist for future discussion. First, we have explored the efficacy of field versus experimental approaches to the study of landscapes. The rewards and pitfalls of each approach are honestly confronted. Just as in other fields, the greatest improvements in understanding in landscape ecology will undoubtedly come from reaping the benefits of both protocols. Second, we have explained how the trophic organization of communities can be profoundly influenced by the landscape context in which they occur and how this in turn determines the population dynamics of constituent species.

Finally, we have exposed two perspectives on the landscape concept itself. On the one hand, and more traditionally, landscapes are limited to units of study that incorporate human spatial scale and human-modified communities, forcing us to search for pattern and understanding in a system context of massive dimensions. Whether or not it makes sense to deny that such units of investigation are appropriately called ecosystems because they are in some sense at a higher organizational level is an issue for debate.

An alternative view that I personally find especially rewarding is to place landscapes at the level of biological organization above that of the community. As such, landscapes express a long-felt but previously not clearly articulated intuition that such a level is important to ecological science. For some, the term *ecosystem* has filled this conceptual niche. Ecosystems, however, are a holistic expression of the systems nature of our units of study, and hence the concept applies to all levels of biological complexity of interest to ecologists (organism and above). The ecosystem has never been defined explicitly to express a new level of biological complexity. Landscapes, on the other hand, with their components of community-types, do add new emergent properties not shown by these parts. Holistic philosophy should, I believe, be applied up and down the biological hierarchy and not be restricted to just communities or just landscapes.

Any book that explores, explains, and exposes so many important topics must be worth our effort and we hope will justify critical reading.

Contributors

Gary W. Barrett, Institute of Ecology, University of Georgia, Athens, GA 30602, U.S.A.

John A. Bissonette, Utah Cooperative Fish and Wildlife Research Unit, Department of Fisheries and Wildlife, Utah State University, Logan, UT 84322, U.S.A.

Sim Broekhuizen, Institute for Forestry and Nature Research, P.B. 9201, 6800 HB Arnhem, The Netherlands

James E. Diffendorfer, Museum of Natural History, University of Kansas, Lawrence, KS 66045, U.S.A.

Michael S. Gaines, Department of Biology, University of Miami, P.O. Box 249228, Coral Gables, FL 33124, U.S.A.

Lennart Hansson, Department of Wildlife Ecology, Swedish University of Agricultural Sciences, Box 7002, S-750 07 Uppsala, Sweden

Steven J. Harper, Department of Ecology, Ethology, and Evolution, University of Illinois, Urbana, IL 61820, U.S.A.

Robert D. Holt, Museum of Natural History, University of Kansas, Lawrence, KS 66045, U.S.A.

Michał Kozakiewicz, Institute of Zoology, Warsaw University, Krakowskie Przedmieście 26/28, 00–927/1 Warsaw, Poland

William F. Laurance, Division of Wildlife and Ecology, Tropical Forest Research Centre, CSIRO, P.O. Box 780, Atherton, Queensland 4883, Australia

William Z. Lidicker Jr., Museum of Vertebrate Zoology, University of California, Berkeley, CA 94720, U.S.A.

H. Gray Merriam, Department of Biology, Carleton University, Ottawa K1S 5B6, Canada

Tarja Oksanen, Department of Animal Ecology, University of Umeå, S-901 87 Umeå, Sweden

John D. Peles, Department of Zoology, Miami University, Oxford, OH 45056, U.S.A.

Michael Schneider, Department of Animal Ecology, University of Umeå, S-901 87 Umeå, Sweden

Norman A. Slade, Museum of Natural History, University of Kansas, Lawrence, KS 66045, U.S.A.

Jakub Szacki, Institute of Physical Planning and Municipal Economy, ul. Krzywickiego 9, 02–078 Warsaw, Poland

Author Index

(This index does not cover the literature cited at the end of each chapter or acknowledgments.)

Abramsky, Z., 178
Addicott, J. F., 69
Akçakaya, H. R., 33
Allen, T. F. H., 157
Alverson, W. S., 161
Ambuel, B., 59, 142
Amores, F., 103
Anderson, P. K., 4–5, 21–2
Andrén, H., 87, 141–2
Andrzejewski, R., 79, 82
Angelstam, P., 33, 87, 141–2
Ansorge, H., 103, 114
Archer, M., 55
Archibald, W. R., 111
Ås, S., 32

Babińska, J., 79
Babińska-Werka, J., 79–80
Balharry, D., 97, 104, 106, 108, 110–2
Barrett, G. W., 33, 155, 157–62, 166–71, 176, 196
Barrett, R. H., 99
Bassett, C. F., 111
Bateman, M. C., 95
Bauchau, V., 81, 86
Baudry, J., 28, 29, 32
Beckon, W. N., 175
Begon, M., 78
Beier, P., 72
Bennett, A. F., 32, 58, 68, 72
Bergerud, A. T., 100
Bergström, M.-R., 145
Bernard, R., 112

Bissonette, J. A., 8, 33, 44, 59, 64, 66–7, 95–9, 101–3, 108, 111, 113–4, 141, 157
Bjärvall, A., 145
Bock, E., 79
Bohlen, P. J., 157
Boitani, L., 144
Bowen, B. S., 29
Bowers, M. A., 154
Brainerd, S. M., 113–5
Brassard, J. A., 112
Breitenmoser, U., 144–5
Breitenmoser-Würsten, C., 145
Bright, P. W., 141
Brinck, P., 21
Broekhuizen, S., 8, 33, 44, 59, 64, 66–7, 97–8, 104, 106, 111–3, 141, 157
Brown, I. L., 176
Brown, J. H., 7, 56, 67, 80
Brown, J. L., 139
Brown, J. S., 79–80
Buchalczyk, T., 79
Buechner, M., 161
Bujalska, G., 25, 87
Bunnell, F. L., 103
Burgin, A. B., 71, 78
Burkey, T. V., 58
Buskirk, S. W., 96, 99, 101–3, 108

Campbell, T. M., 98, 103, 110–1
Carter, S. C., 193
Cary, J. R., 161
Charnov, E. L., 125, 128–9, 143
Ciucci, P., 144

Clark, B. K., 81
Clark, P. J., 8
Clark, T. W., 98, 103, 108, 110–1
Clevenger, A. P., 97
Clough, G. C., 98
Cockburn, A., 5, 9
Cole, F. R., 168
Colgan, P. W., 98, 111
Collins, R. J., 162, 169
Comins, H. N., 66, 140
Council of Europe, 145
Covacevich, J., 55
Covey, S. R., 171
Cox, G. W., 14, 32, 58, 59, 73, 89
Crome, F. H. J., 48, 50, 58
Crowley, P. H., 140
Crowner, A. W., 159

Daily, G. C., 74
Danielson, B. J., 9, 127, 175
Degnn, H. J., 97
Delattre, P., 12–3
Delibes, M., 98, 103
den Boer, P. J., 21
de Rebeira, C. P., 58
Diamond, J. M., 53, 57, 158, 175
Dickman, C. R., 81
Diekmann, O., 140
Diffendorfer, J. E., 25, 33, 155, 158, 169, 193, 195
Doak, E. D., 196
Dobrowolski, K., 88
Dolch, D., 97
Doncaster, C. P., 81, 141
Douglass, C. W., 108
Drake, J. A., 7
Draper, N. M., 186
Drew, G. S., 96–7
Dunning, J. B., 71, 74, 78, 82

Edler, A., 21
Ehrlich, P. R., 74, 176
Ekerholm, P., 136, 139
Erlinge, S., 135, 139
Erwin, T. L., 59
Evans, F. C., 8, 20

Fahrig, L., 21, 28–9, 58, 65–8, 71–3, 79, 88, 158, 167, 175–6
Fenyuk, B. K., 4

Fiedler, P. L., 14, 33
Forman, R. T. T., vii, 3, 113, 141
Foster, J., 25, 176, 178, 194–5
Foster, R. B., 59
Francevic, L. I., 87
Francis, G. R., 108, 111
Frank, F., 26
Franklin, J. F., 60
Frawley, K. J., 47–8
Fredrickson, R. J., 111
Free, C. A., 140
Fretwell, S. D., 122–5, 127, 140
Fuller, K., 110

Gaines, M. S., 24, 25, 33, 128, 155, 176–8, 193–5
Gallagher, P. B., 32, 72, 89
Getz, L. L., 193
Gilpin, M. E., 6, 22, 53, 87, 175
Glass, G. E., 178
Glenn, S. M., 31
Gliwicz, J., 79, 83, 86
Godron, M., vii, 3, 113
Gortat, T., 82–3
Gossow, H., 144
Goszczyński, J., 87
Gotelli, N. J., 68
Goundie, T. R., 70
Grant, J. D., 52
Grant, P. R., 53, 178
Grupe, G., 103
Gyllenberg, M., 140

Haila, Y., 176
Hairston, N. G., 123
Hairston, N. G., Jr., 123
Häkkinen, I., 144
Hall, A. T., 159, 170
Haller, H., 144
Hanski, I., 6, 22, 29, 31, 67, 72, 87, 139–40, 176
Hansson, L., 3–5, 11, 21–2, 25–6, 33, 65–6, 87–9, 127, 132, 141, 158, 175, 195
Harestad, A. S., 103
Hargis, C. D., 96, 113–4
Harper, S. J., 155, 158–9, 161–2, 166, 176
Harris, L. D., 32, 58, 72, 86, 89, 161
Harrison, S., 6, 80, 176
Hassell, M. P., 140
Hastings, A., 140, 175

Hawley, V. D., 101, 103, 108, 110–1, 113
Henderson, M. T., 21, 28, 31, 67, 69, 158, 167, 176
Henein, K., 28, 29, 58, 68–9, 74, 86, 88
Henttonen, H., 33, 124, 139, 141
Heptner, V. G., 97, 113
Herman, T., 110
Herrmann, M., 98, 100, 104, 106, 108, 111, 113
Heske, E. J., 9
Hobbs, R. J., 32, 50, 59, 73, 141
Hoekstra, T. W., 157
Hoffmeyer, I., 86
Holišova, V., 98
Holt, R. D., 123, 125–8, 139–42, 155, 175
Honsig-Erlenburg, P., 144
Hopkins, M. S., 46
Hornocker, M. G., 98
Hudson, W. E., 72
Hughes, R., 46
Hurlbert, S. H., 159, 170–1

Ims, R. A., 33, 66, 154–6, 158, 169

Jacobs, M., 141
Jain, S. K., 14, 33
Janzen, D. H., 51
Jensen, B., 97
Jessop, R. H., 111
Joseph, L., 48
Judson, O. P., 68
Jurasińska, E., 88

Kalpers, J., 103
Kangas, P., 86, 89
Kapos, V., 51
Kareiva, P., 140, 176
Kelly, W. G., 68
Kemp, J. C., 171
Kershaw, A. P., 46
Keto, A., 46
Kilgore, D. L., 99
Klosterman, L. L., 9, 25
Knaapen, J. P., 28
Kodric-Brown, A., 7, 56, 67
Koehler, G. M., 96, 98
Kolasa, J., 64, 78
Konopka, J., 29
Koshkina, T. V., 4
Kozakiewicz, A., 79

Kozakiewicz, M., 11, 29, 43, 58, 67, 69, 80, 82–4, 88–9, 158, 195
Krebs, C. J., 32, 180
Krebs, J. R., 129, 141
Krohne, D. T., 9, 71, 78
Krott, P., 112
Kruger, H. H., 103
Kruuk, H., 141
Kuitunen, J., 31

Lachat Feller, N., 98, 100, 103
Lammertsma, D., 100
Lande, R., 175
Lankester, K., 11, 15, 32, 68, 72
Lanoue, A., 28, 72, 88, 168
La Polla, V. N., 158, 160, 162, 168
Larsson, T.-B., 26
Laudenslayer, W. F., Jr., 161
Laurance, W. F., 11, 32, 43, 46–53, 55–60, 66, 72, 158, 161, 176, 196
Le Boulengé, E., 81, 86
Lefkovitch, L. P., 21, 28
Lensink, C. J., 101, 111
Leopold, A., 3
Leung, L. K. P., 48
Levins, R. A., 6, 21, 26, 130
Lidicker, L. N., ix, 16
Lidicker, W. Z., Jr., 3, 4, 8–9, 11, 14, 24, 33, 60, 78–81, 87, 115, 122, 127–8, 132, 153, 155, 157–8, 175, 196
Lieberman, A. S., 3
Likens, G. E., 159
Lindström, E., 141
Liro, A., 70, 81, 83–4, 89
Łomnicki, A., 127–8
Lomolino, M. V., 22, 29
Lord, J. M., 78, 175
Lorenz, G. C., 158, 160, 168
Lovejoy, T. E., 51, 176
Lubchenco, J., 170–1
Lucas, H. L., 125, 127, 140

MacArthur, R. H., 7, 21, 127–8, 130
Mader, H., J., 29
Madsen, A. B., 103
Major, J. T., 96, 98–9, 101, 110
Maly, M. S., 166
Marchesi, P., 97–8, 100, 103–5, 114
Markley, M. H., 111
Martell, A. M., 98

Martin, S. K., 96, 99
Martinsson, B. L., 142
McCallum, H. I., 175
McCullough, D. R., 96
McDonald, L. L., 103
McNaughton, S. J., 123
Mech, L. D., 141
Meldžiūnaite, S., 112
Mermod, C., 103
Merriam, G., vii, 11, 21–2, 28–9, 31–2, 43,
    56, 58, 64–72, 74, 78–9, 81, 88, 113, 158,
    167–8, 175–6
Middleton, J. D., 22, 28, 31, 65, 71
Miller, R. G., 111
Milne, B. T., 78
Moen, J., 137
Monthey, R. W., 98
Moritz, C., 48
Morris, D. W., 9, 70, 79, 127–8, 140
Morrison, M. L., 9
Mrciak, M., 21
Murie, A., 101
Müskens, G. J. D., 97–8, 100, 104, 106, 111
Myllymäki, A., 5

Nagorsen, D. W., 102
Nams, V. O., 8
Naumov, N. P., 4, 20, 21
Naveh, Z., 3
Nesvadbova, J., 97
Newby, F. E., 102, 110–1, 113
Newsome, A. E., 26
Nicholson, A. J., 70
Nicht, M., 98
Norton, D. A., 78, 175
Noss, R. F., 32
Nudds, T. D., 31
Nyholm, E. S., 111

Obrtel, R., 98
Odum, E. P., 169–70
Oksanen, L., 122–5, 127–30, 133, 136, 139
Oksanen, T., 10, 14, 33, 44, 87, 115, 122,
    124–5, 128, 130–3, 135–7, 139–44, 158,
    195
O'Neill, R. V., 158
Ostfeld, R. S., 9, 11–2, 24–5, 122, 125, 128,
    154, 175
O'Sullivan, P. J., 97, 103, 114
Osunkoya, O. O., 52–3

Pacala, S. W., 140
Pahl, L. K., 48
Paloheimo, J., 66, 73, 175–6
Panteleyev, P. A., 4
Paquet, P. C., 70
Patton, J. L., 79, 81
Pavlovic, N. B., 79–80
Pavuk, D. M., 171
Peles, J. D., 155, 162, 167
Pelikán, J., 97
Peltonen, A., 31, 176
Pickett, S. T. A., 64, 113
Pierotti, R., 195
Pimm, S. L., 33
Polyakov, I. Ya., 4
Powell, R. A., 108
Price, M. V., 113
Primack, R. B., 15
Promberger, C., 145
Pucek, Z., 79
Pulliainen, E., 97, 111, 141
Pulliam, H. R., 9, 115, 125, 127–8, 175
Purves, H. D., 70

Quinn, J. F., 175

Radvanyi, A., 98
Raine, R. M., 96
Rasmussen, A. M., 98, 103
Raup, D. M., 15
Ray, C., 28
Redfield, J. A., 178
Redhead, T. D., 26, 28, 33
Reeve, J. D., 140
Risser, P. G., 3, 21, 22, 157, 159
Robinson, G. R., 158, 170
Roff, D. A., 175
Rollo, C. D., 78
Rosenzweig, M. L., 124–5, 127–8, 139
Rzebik-Kowalska, B., 103, 114

Sabelis, M. W., 140
Sandell, M., 139
Saunders, D. A., 32, 50, 58, 141
Schaller, G. B., 141
Schmidt, F., 112–3
Schneider, M., 10, 14, 33, 44, 87, 115, 144–5,
    158, 195
Schreiber, K. F., 3
Schröder, W., 145

Schröpfer, R., 105
Scott, K., 46
Sheppe, W., 79
Sherburne, S. S., 96–7, 99, 103
Shvarts, S. S., 4
Sietman, B. E., 195
Simberloff, D., 32, 58–9, 73, 89, 175
Simon, T. L., 99
Sinclair, A. R. E., 123
Singleton, G. R., 28, 33
Skirnisson, K., 98, 103–4, 106, 113
Slade, N. A., 155, 168, 178–9
Slough, B. G., 101
Smith, A. T., 28
Smith, H., 186
Smith, M. H., 5
Sonerud, G. A., 141
Soulé, M. E., 14, 31, 176
Southwood, T. R. E., 65, 79, 82, 83
Soutiere, E. C., 98, 101
Spencer, W. D., 96, 99–100
Ssemakula, J., 22, 31
Stamps, J. A., 161, 166
Starr, T. B., 157
Steiner, H. M., 81
Stenseth, N. C., 6, 21–2, 24–6, 33, 66, 78–80, 154–6, 158, 169
Stephens, D. W., 129
Stephenson, A. B., 98, 108, 111
Steventon, J. D., 98–9, 101
Stickel, L. F., 21
Storch, I., 97, 103, 105
Strickland, M. A., 101, 108
Stueck, K. L., 159, 166
Swihart, R. K., 168
Szacki, J., 11, 29, 43, 58, 67-70, 80–1, 83–4, 88–9, 158, 195

Taitt, M. J., 32
Taylor, A. D., 140
Taylor, P. D., 28, 66, 71, 74
Temple, S. A., 59, 142, 161
Terborgh, J., 57
Terman, M. R., 178
Tester, U., 98, 103
Tew, T., 81
Thomas, J. W., 15
Thompson, I. D., 98, 101, 103, 111
Threader, R. W., 100
Tracey, J. G., 47–8, 59

Tucker, B. M., 101
Turner, M. G., 3, 64, 160

Urban, D. L., 157–8
Usher, M. B., 11

Vačkař, J., 97
van Apeldoorn, R. C., 11, 29, 32, 173, 176
Van Bostelen, A. J., 100
Van Dyck, S., 53
Van Horne, B., 9, 115
Verboom, J., 11, 32, 68
Verhoog, M. D., 100
Vessey, S. H., 70
Viitala, J., 86
von Somsook, S., 81

Warner, P., 103, 114
Webb, L. J., 46
Weckwerth, R. P., 101, 103
Wegner, J. F., 21–2, 29, 31, 65, 69–70, 74, 81, 86, 158, 167, 176
Western, D., 22, 31
Wharburton, N., 48
White, P. S., 113
Wielgolaski, F. E., 124
Wiens, J. A., 4, 8, 78–9, 113, 141
Wilcove, D. S., 169
Wilcox, B. A., 57
Williams, C. K., 171
Willson, M. F., 50
Wilson, D. S., 7
Wilson, E. O., 7, 15, 21
Wilson, P., 195
Winter, J. W., 47–8
Wolff, J. O., 9
Wolton, R. J., 81
Woods, J. G., 72
Worthen, G. L., 99
Wynne, K. M., 99

Yahner, R. H., 161
Yensen, E., 48

Zejda, J., 97
Zhang, Z., 11
Zielinski, W. J., 111
Zimen, E., 144
Zonneveld, I. S., 3, 141

Subject Index

Abies (fir), 100
  balsamea, 95
Africa, 31
Alphitonia petrei, 56
Antechinus, 53
  flavipes (yellow-footed), 53–4
  godmani (Atherton), 53
  stuartii (brown), 53
anthropogenic index, 6
Apodemus, 86
  agrarius, 81
  flavicollis, 88
  sylvaticus, 81, 86
Australia, vii, 26, 32, 34, 46, 200
  Queensland, Chap. 3, 34

badger. See *Meles*
bandicoot
  brown, 55
  long-nosed, 55
Betula, 122
  nana, 133
  papyrifera, 95
  pubescens, 133
biogeography, island, 7, 21–2, 28, 31–4, 69,
  175
birch. See *Betula*
Bison bison, 123
Bonasa umbellus, 103
Bromus, 179
Bufo marinus (cane toad), 55, 59

Calamus, 48, 50–1
Canada, 21, 31, 34. *See also* North America

Banff National Park, 70
British Columbia, 102
Jasper Park, 71
Newfoundland, 95–7, 101–2, 111
Ontario, 29, 86
Yukon, 71
Canis. See also dog
  latrans, 97
  lupus, 70, 144
Capreolus capreolus (roe deer), 144
carrying capacity, 10, 24, 87, 130, 193
cat. See *Felis*
cattle, 59
Cervus elaphus (red deer), 144
chipmunk. See *Tamias*
Circaeo-Alnetum forest type, 82
Clethrionomys, 86, 102
  glareolus, 25, 29, 80–3, 86, 88
  rufocanus, 136
  rutilus, 136
coaction, 14, 43
community-type, 4, 6–8, 14–6, 153, 202
competition, 123, 127, 142
  interspecific, 53, 177–8, 193–4, 196–7
  intraspecific, 104, 116, 192, 194
connectivity. *See* ecology, landscape: connectedness
conservation, 3, 14–6, 21, 31–4, 46, 88–90,
  95, 113, 122, 141–5, 175, 196, 201
  biodiversity, vii, 14, 16, 69, 89, 175,
    201
  reserves, 31–2, 46, 48, 58–60, 69,
    175
corridor, 11, 15, 21, 23, 28–9, 31–2, 34, 43,

48, 56–60, 72–4, 88–90, 155, 159–60,
    167–9, 172
  design, 58–60, 72–3, 143–5
coyote, 97
Cretaceous period, 1
cycles, multiannual, 10–2, 24–26, 136–7,
    139, 181

Dasyurus maculatus, 52, 55, 57
deer. See *Capreolus; Cervus; Odocoileus*
deme, 6, 15, 88
demographic unit, operational (ODU), 28,
    43, 65, 68–70, 75
Dendrocnide, 50–1
Dendrolagus, 43
    lumholtzi, 55–6
dispersal, 6–11, 15, 20, 22–4, 28, 31, 34, 43,
    56–8, Chap. 4, Chap. 5, 111–3, 115,
    127–8, 159–61, 166, 168, 172, 190, 192–3,
    195. *See also* ecology, landscape
  evolution of, 34, 64–7, 73–4
  frustrated, 12, 24
  insect, 171
  landscape movement, 65–6, 73–5,
    80–4
  measuring, 79–80
  over ice, 29
  seed, 59
dispersion, 5–6, 65
diversity, community, 6
dog, 59
Dryocopus martius (black woodpecker), 100
dunnart, 53

ecology, landscape, Chap. 1, 21–2, 32–4,
    43–4, 53, 60, 74, 113, 141–3, 145, 153–4,
    157–8, 160–1, 169–71, 196, 201. *See also*
    landscape
  colonization, 6–7, 22, 31, 67–8, 87
  connectedness, 6, 8, 28–31, 34, 58, 66,
    68, 70–3, 75, 88, 160, 167–9
  edge-to-area ratios, 6, 8, 34, 159, 161,
    166, 192, 194
  extinction, 6–8, 10, 12, 14, 22–3, 31, 43,
    53, 56, 67, 74, 87, 140, 143, 145, 167,
    175 (*see also* extinction proneness)
  history, 3–5, 20–2, 157–8, 175–6
  matrix, 6–7, 10–2, 25, 28–32, 34, 43,
    52–3, 56–8, 60, 73, 90, 125, 140–1,
    159, 168, 172, 194–5

species richness (diversity), 58–60, 175
  survival strategies, 83–6, 90
ecosystem, 64, 122, 141, 155, 157, 159, 166,
    169–70, 201–2
  exploitation, 122–3, 132–3, 145
edge effects, 6, 15, 48, 50–2, 55, 58–60, 90,
    115, 141, 161
edge-to-area ratios. *See* ecology, landscape:
    edge-to-area ratios
emergent properties, 4, 6, 155, 202
Empetrum hermaphroditum (crowberry),
    133
epiphyte, 48
Europe, 3, 15, 32–4, 44, 101, 145. *See also*
    Soviet Union
  Austria, 144
  Bavaria, 102
  Belgium, 86
  Czech Republic, 97
  Denmark, 97, 102
  Finland, 97, 111, 133, 139, 143–4
  France, 12–3
  Germany, 26, 102, 106 (*see also* Europe:
    Bavaria, Schleswig-Holstein)
  Great Britain, 20 (*see also* Europe: Scot-
    land, Welsh island)
  Ireland, 97
  Italy, 144
  Netherlands, The, 15, 29, 72, 104, 107–9,
    114
  Norway, 44–5, 133–9, 143, 145, 154–6
  Poland, Kampinos Forest, 82–3
  Scandinavia, 3, 21, 34 (*see also* Europe:
    Denmark, Finland, Norway, Sweden)
  Schleswig–Holstein, 102
  Scotland, 97, 104, 111
  Spain, Minorca, 97
  Sweden, 21, 133, 144
  Switzerland, 144
    Jura Mountains, 97, 105
  Welsh island, 81
extinction proneness, 56–7. *See also* ecol-
    ogy, landscape: extinction

Felis, 59
    concolor coryi, 32, 72
fir, 95, 100
fox, 33
    red, 97
frugivore, 59

gene flow, 4, 68, 88
genetics, 8, 14, 48, 56, 70, 79, 88, 159–60
   founder effects, 29
Geographic Information System (GIS), 74
gerbil, 20
grass. See *Bromus; Oplismenus; Panicum*
grouse, ruffed, 103
Gulo gulo (wolverine), 145

habitat-type (patch), 10–4, 20, 27, 29, 78, 88,
   122, 129, 132, 145, 159, 172, 176, 196–7
   juxtaposition, 9–11, 15–6, 33, 124–5
   key (critical), 89–90
   quality, 3–5, 9–12, 15–6, 24–5, 27, 34, 84,
   114–5, 124–5, 128, 139, 145, 159, 161,
   166–7, 172
   selection of, 125–31, 140, 145
hare, 33, 111
Hemibelideus lemuroides, 55–7
hierarchy. *See* levels of organization
holism, 4, 14–5, 170, 173, 201–2
home range, concepts of, 44, 86–8, 98–9,
   115, 138, 143, 166, 169
Hubbard Brook (experimental forest), 159
Hydromys chrysogaster, 53
Hypsiprimnodon moschatus, 53

island biogeography. *See* biogeography, island
Isoodon macrourus, 55

Jolly-Seber models, 180, 187

keystone species, viii

landscape, vii, 3, 5–7, 10, 15–6, 20, 30, 45,
   52, 64–5, 78–9, 130, 134, 153, 155–6,
   166–7, 169, 172, 201–2. *See also* ecology,
   landscape
   composition, 20, 22–32, 65, 70–1, 75, 78,
   166–7, 172
   configuration, 66, 70–1, 75, 166–7, 172,
   196
   definition, 4–5, 7, 15, 64, 157, 201
   experimental, 25, 33, 153–6, Chap. 8,
   Chap. 9, 201
   physiognomy, 78
lemming, 26, 144. See also *Lemmus*
Lemmus lemmus, 137, 139. *See also* lemming

Lepus (hare), 33
   americanus, 111
levels of organization, 4, 7, 15, 153–5,
   157–8, 160, 172, 201–2
liana, 50–1. *See also* vine
lichen, 133, 136
Lotka-Volterra models, 139
Lynx lynx, 144

marsupial, 32, 43, 50, 54–5
marten. See *Martes*
   American (see *M. americana*)
   European pine (see *M. martes*)
   stone (see *M. foina*)
Martes, 44, Chap. 6
   americana, 95–7
   diet, 101–3
   foina, 64, 98
   home range, 98–9, 103–13, 115
   local movements, 111–2
   martes, 97–8
   pennanti (fisher), 96
   population density, 100–1
   resting sites, 99–100
matrix. *See* ecology, landscape: matrix
Meles, 32
   meles, 15, 72
Melomys cervinipes, 51–2
mesocosm (mesoscale), 153, 159, 169–70,
   172, 175
metacommunity, 5, 7, 15
metapopulation, 5–7, 11, 15, 22, 26, 28–34,
   43, 86–8, 140, 143, 191
   Levins's classic, 6–7, 26
   nonequilibrium, 7
   patchy populations, 6, 87
Microtus, 102
   agrestis, 136
   arvalis, 12–3, 81
   californicus, 9, 25, 29
   ochrogaster, 176, 178–81, 183–4, 186–91,
   193–7
   oeconomus, 136, 155–6
   pennsylvanicus, 101–2, 111, 161, 166–9
   xanthognathus, 81
Morelia (python), 52
mouse, house. See *Mus*
Mus, 22, 26
   musculus, 4, 21, 55, 81, 166–8
Mustela, 33

erminea, 135–9, 143–4
    nivalis, 12, 124, 135–9

Napeozapus, 102
nomad, 83–4, 87, 90
North America, 3, 15, 21, 29, 44, 86, 101–2,
    106, 170. *See also* Canda; United States
    Rocky Mountains, 71, 99

Ochotona princeps (pika), 28
Odocoileus virginianus (white-tailed deer),
    144
Omalanthus novo-guineensis, 59
Oplismenus hirtellus, 50
owl, 15, 52

pademelon, 50
Panicum, 9
panther (cougar), Florida. See *Felis
    concolor*
parasite (disease), 8, 10, 20–1, 23, 26
patch. *See* habitat-type (patch)
Perameles nasuta, 55
Peromyscus, 98, 102
    leucopus, 28–9, 74, 81, 86, 167
    maniculatus, 25, 176, 178–81, 183,
        185–8, 190–1, 194–7
pest control, 21, 23, 26, 28, 33–4
Phenacomys, 102
phytosociology, 3
PICA (strategy), 28
Picea (spruce), 95, 100
pig, feral, 59
pika, 28
Pinus contorta (lodgepole pine), 96–7, 100
Plectrophenax nivalis, 124
Pleistocene period, 46
population dynamics, multifactor, 33–4, 141
possum. See *Hemibelideus; Pseudocheirops;
    Pseudocheirus; Trichosurus*
Precambrian period, 133
predator, 8, 10–2, 20, 22–3, 25–6, 33–4, 44,
    52–3, 55, 59, 87, 96–7, 115, Chap. 7. See
    also *Canis; Dasyurus; Felis; Martes;
    Meles; Morelia; Mustela*
    effects on prey density, 123–33, 144–5,
        195 (*see also* trophic exploitation hy-
        pothesis)
    eggs, 53
    insects and small vertebrates, 52

seeds, fruits, 52–3
spillover, 126–7, 130–2, 139, 145
    prediction, 65, 67–75, 122, 124, 139,
        153–4, 161, 179
prey. *See* predator
Pseudocheirops archeri, 54–7
Pseudocheirus herbertensis, 54–6
python, 52

quoll. See *Dasyurus*

rainforest, Chap. 3
Rangifer tarandus (reindeer), 144–5
rarity, responses to fragmentation, 53, 57
rat
    black (see *Rattus rattus*)
    bush (see *Rattus fuscipes*)
    cane (see *Rattus sordidus*)
    Cape York (see *Rattus leucopus*)
    cotton (see *Sigmodon*)
    swamp (see *Rattus lutreolus*)
    water (see *Hydromys*)
    white-tailed (see *Uromys*)
rat-kangaroo, musky, 53
rattan, climbing. See *Calamus*
Rattus, 51
    fuscipes, 51–3
    leucopus, 51–3
    lutreolus, 55
    rattus, 55
    sordidus, 55
refugia, 26, 46. *See also* habitat–type
reindeer, 144–5
Reithrodontomys megalotis, 81
rescue effect, 7, 56, 67
ROMPA (hypothesis), 11–2, 24–5, 27, 128,
    132
Rubus, 9, 10
    alcefolius, 50
    idaeus (raspberry), 96, 98
Rupricapra rupricapra (chamois), 144
Russia. *See* Soviet Union

Salici–Franguletum forest type, 82
Salix (willow), 124, 133
Sciurus, 32
seutu, 5, 7, 16, 155
sex ratio, 9, 68, 176, 180, 188, 193, 196
shrew, 26. See also *Sorex*
Sigmodon hispidus, 9, 176–83, 186–97

sink, 9–11, 24, 67, 86, 115, 127
Sminthopsis leucopus, 53
social behavior, 8, 25, 34, 70, 87, 106–7, 127, 156, 192–5
Solanum, 51
  dallachii, 50
  hamulosum, 50
  mauritianum, 50
Sorex, 29, 102
South America, 43
Soviet Union, 3, 4, 20–1, 33. *See also* Europe
spatial scale, effects of, 8, 64–5, 70, 74, 78, 95, 113, 115–6, 153, 158, 160, 175, 196
spruce, 95, 100
squirrel, 102–3. See also *Sciurus; Tamiasciurus*
stability, 6, 12, 14, 21
steppe, 20, 124
stinging trees (*Dendrocnide*), 50–1
Strix occidentalis (Spotted Owl), 15
Synaptomys, 102

Tamiasciurus hudsonicus, 99
Tamias striatus, 68, 74, 86, 167
temporal scale, 79, 158, 160
Theriological Congress, Sixth International, vii
Thomomys bottae, 81
Thylogale stigmatica, 50
toad, cane, 55, 59
treefall gaps, 50
tree-kangaroo. See *Dendrolagus*
Trichosurus vulpecula, 55–6
Trifolium, 155
trophic exploitation hypothesis, 12, 44, 139,

Chap. 7. *See also* ecosystem, exploitation
tundra, 122, 124, 133, 136

United States of America, 34. *See also* North America
  Alaska, 96, 111
  California, 96, 99
  Idaho, 96
  Kansas, 123, 155, 176–95
  Maine, 98–9, 101
  New York, 154
  Ohio, 155, 161
  South Carolina, 9
  Utah, 114
  Virginia, 154
  Wyoming, 96
  Yellowstone National Park, 70, 97
Uromys caudimaculatus, 52

Vaccinium myrtillus (blueberry), 133
vine, notophyll, 48. *See also* liana
vole, 22, 26, 139, 144. See also *Clethrionomys; Microtus; Phenacomys; Synaptomys*
Vulpes vulpes, 97. *See also* fox

wallaby, 50
weasel. See *Mustela*
weed, 59, 73
willow (*Salix*), 124, 133
witches'-broom, 100
wolf, 70, 144
wolverine, 145
woodpecker, black, 100

Zapus, 102